Annals of Mathematics Studies

Number 140

Global Surgery Formula for the Casson-Walker Invariant

by

Christine Lescop

PRINCETON UNIVERSITY PRESS

PRINCETON, NEW JERSEY

1996

Princeton University Press books are printed on acid-free paper and meet the guidelines for permanence and durability of the Committee on Production Guidelines for Book Longevity of the Council on Library Resources

Printed in the United States of America by Princeton Academic Press

10 9 8 7 6 5 4 3 2 1

Library of Congress Cataloging-in-Publication Data

Lescop, Christine, 1966–
Global surgery formula for the Casson-Walker invariant / by Christine Lescop.
p. cm. — (Annals of mathematics studies ; no. 140)
Includes bibliographical references and index.
ISBN 0-691-02133-3 (cloth : alk. paper). — ISBN 0-691-02132-5 (pbk. : alk. paper)
1. Surgery (Topology) 2. Three-manifolds (Topology) I. Title. II. Series.
QA613.658.L47 1996 95-45797
514'.72—dc20

The publisher would like to acknowledge the author of this volume for providing the camera-ready copy from which this book was printed

Global Surgery Formula for the Casson-Walker Invariant

Table of contents

Chapter 1

Introduction and statements of the results

§1.1 Introduction

In 1985, A. Casson defined an integer invariant for oriented integral homology 3-spheres by introducing an appropriate way of counting the conjugacy classes of the SU(2)-representations of their fundamental group. He proved that his invariant λ_c satisfies the following interesting properties :

λ_c vanishes on homotopy spheres,

λ_c is additive under connected sum,

λ_c changes sign under orientation reversal,

a simple surgery formula describes the variation of λ_c under a surgery on a knot transforming an integral homology sphere into another one, and,

λ_c lifts the Rohlin μ-invariant from $\mathbb{Z}/2\mathbb{Z}$ to $\mathbb{Z}$ (recall that if M is a $\mathbb{Z}$-homology 3-sphere, $8\mu(M)$ is the signature (mod 16) of any smooth spin 4-manifold with boundary M).

This last property allowed Casson to answer several old and well-known questions about the Rohlin invariant and the topology of 3-manifolds (see [G-M2] or [A-M]).

In 1988, K. Walker extended the Casson invariant, and all of its original properties, to rational homology 3-spheres; furthermore, he gave a combinatorial and elementary definition for his extension (see [W]).

According to a theorem independently proved by Lickorish and Wallace (see [Rou]), any compact connected oriented 3-manifold without boundary can be presented by a surgery diagram (or a framed link) in S^3. A theorem of Kirby describes simple moves which suffice to relate two surgery presentations of the same 3-manifold (see [Kir 2]).

This book states and proves a global surgery formula for the Casson-Walker invariant, that is it describes explicitly a function $\mathbb{F}$ of the surgery diagrams

presenting rational homology spheres such that $\mathbb{F}$ gives the Casson-Walker invariant of the manifolds presented by such diagrams.
This function $\mathbb{F}$ extends naturally to all surgery diagrams in S^3. In Chapter 3, it is verified directly to be invariant under the Kirby moves; it therefore defines an invariant λ of closed oriented 3-manifolds.
The function $\mathbb{F}$ also extends to surgery diagrams in any rational homology sphere and provides, as shown in Chapter 4, a surgery formula describing the variation of λ under any surgery starting from a rational homology sphere. This surgery formula generalizes the Walker one-component surgery formula.
Chapter 5 describes the invariant λ as a function of previously known invariants for manifolds with nonzero first Betti number. λ becomes simpler as the first Betti number increases, vanishing for manifolds with first Betti number greater than 3.
These results are the main results of this book. They are precisely stated in §1.5, after the description of $\mathbb{F}$ in §1.4, which involves notation introduced in §1.2 and §1.3. Section 1.6 outlines the proofs of these results and refers to the following chapters for details.

The function $\mathbb{F}$ is defined for any rational surgery presentation in any rational homology sphere. In the case of an integral surgery presentation, the signature of the associated 4-dimensional cobordism is part of the function $\mathbb{F}$. This yields a straightforward comparison (Proposition 6.3.8) between λ and the Rohlin μ-invariant; this also allows us to give a new proof that μ is well-defined in §6.3.C.
§6.1 applies the surgery formula to the computation of λ for all oriented Seifert fibered spaces, as an example.

§1.4 describes $\mathbb{F}$ as a sum of a combination D of certain derivatives of several variable Alexander polynomials, and a function of linking numbers associated with the presentations.
The part D of $\mathbb{F}$ had been found by S. Boyer and D. Lines in [B-L 1], where they showed in particular that the function of surgery presentations of integral homology spheres ($\lambda_c \circ \chi$ - D) depends only on the homotopical type of the link and on its framing.
(Here χ denotes the function mapping a surgery presentation to its associated 3-manifold.)

§1.7 gives two more definitions for $\mathbb{F}$. Definition 1.7.8 describes $\mathbb{F}$, for surgery presentations with null-homologous components, as a function of one-variable Alexander polynomials and linking numbers of the presentations.

The corresponding surgery formula generalizes the Hoste surgery formula for the variation of the Casson invariant under surgeries with diagonal linking matrices (see [Hos]).

The main (and only) tool in this book is the normalized several-variable Alexander polynomial. All its required properties are stated in Chapter 2 and proved in the appendix.

Acknowledgements

I am grateful to Steven BOYER, Daniel LINES and Kevin WALKER whose articles [B-L 1] and [W] inspired this one.
I thank Christian BLANCHET, Michel BOILEAU, Lucien GUILLOU, Nathan HABEGGER, Pierre VOGEL and especially Alexis MARIN. They had enough courage to start reading the first version of this book and their remarks were of much use in the rewriting process.
My thanks also go to Viviane BALADI, Nathalie HUNTER-MANDON and Lisa RAMIG for their generous help with my English.
Part of this work has been written when I was a visitor at "l'Université du Québec à Montréal". I warmly acknowledge the hospitality of this University and last but not least the kindness of Steven BOYER during my stay there.

I also thank the referees for their thorough reading of this book and for their appropriate suggestions.

§1.2 Conventions

• The boundary of an oriented manifold is oriented with the "outward normal first" convention; unless otherwise specified, when a manifold and its boundary are oriented, the orientations are supposed to agree.

• An integral (respectively rational) homology sphere is a closed 3-manifold with the same $\mathbb{Z}$-homology (respectively $\mathbb{Q}$-homology) as the usual 3-sphere S^3.

• Curves or surfaces are identified with their homology classes when it does not seem to cause confusion.

• In any closed oriented 3-manifold M, the *linking number* $Lk_M(J,K)$ of two disjoint oriented links J and K representing 0 in $H_1(M;\mathbb{Q})$ is defined as follows:
If J represents 0 in $H_1(M) = H_1(M;\mathbb{Z})$ and if Σ_J is an oriented Seifert surface with boundary J, then $Lk_M(J,K)$ is the algebraic intersection number of Σ_J and K in M. $Lk_M(.,K)$ is next extended by $\mathbb{Q}$-linearity on $Ker(H_1(M\backslash K;\mathbb{Q}) \rightarrow H_1(M;\mathbb{Q}))$.
The linking number $Lk_M(.,.)$ is symmetric.

• The oriented meridian of an oriented knot K in an oriented 3-manifold is the meridian m(K) of K which links K positively.

• If Γ is an abelian group, $|\Gamma|$ denotes its order. (If Γ is finite, $|\Gamma|$ is its cardinality, otherwise $|\Gamma|$ is zero.)

• If K is a knot in a rational homology sphere M, $O_M(K)$ denotes the order of the class of K in $H_1(M)$:

$$O_M(K) = \frac{|\, H_1(M) \,|}{|\, Torsion(\, H_1(M\backslash K)\,)\,|}$$

• Let x be an element of $\mathbb{R}$, $sign(x) = \frac{x}{|x|} = \pm 1$ if x is nonzero and $sign(0) = 0$.

§1.3 Surgery presentations and associated functions

DEFINITION **1.3.1**:
A *primitive satellite of* K is an oriented simple nonseparating closed curve of the boundary $\partial T(K)$ of a tubular neighborhood T(K) of K.
If K is an oriented knot embedded in a rational homology sphere M and if μ is a primitive satellite of K, the homology class $[\mu]$ of μ in $\partial T(K)$ will be identified with the ordered pair (p,q) of $\mathbb{Q}\times\mathbb{Z}$, where

• the curve μ is homologous to qK in T(K),

• $p = Lk_M(\mu,K)$, (p must then be congruent to $qLk_M(K,K)$ modulo $\mathbb{Z}$, where $Lk_M(K,K) \in \frac{\mathbb{Q}}{\mathbb{Z}}$ denotes the self-linking number of K in M, that is the linking number of K and a parallel of K mod $\mathbb{Z}$).

If μ is not a meridian of K, $\pm[\mu]$ will be identified with the rational number $\frac{p}{q}$.

DEFINITION **1.3.2**: *(Surgery presentation)*
A *surgery presentation* in an oriented rational homology sphere M is a link L, each component K_i of which is oriented and equipped with a primitive satellite μ_i, specified by a pair (p_i, q_i) with a positive q_i.
The manifold presented by such a surgery presentation

$$\mathbb{L} = (K_i , \mu_i)_{i=1, \dots, n} = (K_i , (p_i, q_i))_{i=1, \dots, n} = (K_i , \frac{p_i}{q_i})_{i=1, \dots, n}$$

is denoted by $\chi(\mathbb{L})$ and defined in the following way:
Let $T(K_i)$ be a tubular neighborhood of K_i in M, and let h_i be a homeomorphism from the boundary $(S^1 \times S^1)_i$ of $(D^2 \times S^1)_i$ to the boundary $\partial T(K_i)$ of $T(K_i)$ which sends the meridian $(S^1 \times \{*\})_i$ to the curve μ_i of $\partial T(K_i)$, then $\chi(\mathbb{L})$ is the following closed 3-manifold:

$$\chi(\mathbb{L}) = [M \setminus (\coprod_{i=1}^{n} \mathring{T}(K_i))] \cup_{\coprod h_i} [\coprod_{i=1}^{n} (D^2 \times S^1)_i]$$

The surgered manifold $\chi(\mathbb{L})$ inherits the orientation of M.

$\mathbb{L}$ is said to be an *integral surgery presentation* if all the q_i are equal to 1.
The curve μ_i is called the *characteristic curve* of the surgery on K_i and $(\{0\} \times S^1)_i$ is called the *core* of the surgery performed on K_i. (The characteristic curve μ_i is a meridian of the core of the surgery performed on K_i in the surgered manifold.)

DEFINITIONS **1.3.3**: *(Some functions of the surgery presentations)*
Let $\mathbb{L}$ be a surgery presentation in a rational homology sphere M as in 1.3.2.
Let ℓ_{ij} be defined by:

$$\ell_{ij} = \frac{1}{q_j} Lk_M(K_i,\mu_j) = \begin{cases} Lk_M(K_i,K_j) & \text{if } i \neq j \\ \frac{p_i}{q_i} & \text{if } i = j \end{cases}$$

• The *linking matrix* of $\mathbb{L}$ is defined by

$$E(\mathbb{L}) = [\ell_{ij}]_{1 \le i,j \le n}$$

• Note that the matrix $F(\mathbb{L}) = [q_j \ell_{ij}]_{1 \le i,j \le n}$ is a presentation matrix of $H_1(\chi(\mathbb{L}))$ if M is an integral homology sphere, and that, in any case:

1.3.4 $\quad |H_1(M)| \, |\det(F(\mathbb{L}))| = |H_1(\chi(\mathbb{L}))|$

$F(\mathbb{L})$ is called the *framing matrix* of the presentation $\mathbb{L}$.

• The *sign of the surgery presentation* $\mathbb{L}$, denoted by $\text{sign}(\mathbb{L})$, is equal to $(-1)^{b_-(\mathbb{L})}$ where $b_-(\mathbb{L})$ denotes the number of negative eigenvalues of the symmetric matrix $E(\mathbb{L})$.

Note the following equality coming from 1.3.4,

$$\textbf{1.3.5} \qquad \text{sign}(\mathbb{L})\det(E(\mathbb{L}))\prod_{i=1}^{n} q_i = \frac{|H_1(\chi(\mathbb{L}))|}{|H_1(M)|}$$

• We will denote by $\text{signature}(E(\mathbb{L}))$ the signature of $E(\mathbb{L})$ (that is, $(b_+(\mathbb{L}) - b_-(\mathbb{L}))$, where $b_+(\mathbb{L})$ denotes the number of positive eigenvalues of $E(\mathbb{L})$).

• *Restriction of a surgery presentation*

If I is a subset of $N=\{1, \ldots, n\}$, then $\mathbb{L}_I$ (respectively L_I) denotes the surgery presentation obtained from $\mathbb{L}$ (respectively the link obtained from L) by forgetting the components whose subscripts do not belong to I.

Unless otherwise specified, $\mathbb{L}$ will denote the surgery presentation of Definition 1.3.2 and we will use the notation introduced in §1.3 for $\mathbb{L}$.

§1.4 Introduction of the surgery formula $\mathbb{F}$

The *Alexander series* $\mathcal{D}$ is, up to a change of variables (see 2.1.1), the *normalized Alexander polynomial in several variables* Δ, and it will be precisely defined in Chapter 2 as an invariant of oriented links in oriented rational homology spheres.

$$\mathcal{D}(L = \{K_i\}_{i\in\{1,\ldots,n\}}) \in \mathbb{Z}[\exp(\pm\frac{u_i}{2O_M(K_i)})]_{i\in\{1,\ldots,n\}}$$

The surgery formula will only require the following coefficient of the Alexander series:

DEFINITION **1.4.1**: *The ζ-coefficient of a link in an oriented rational homology sphere*

Let $L = (K_i)_{i\in\{1,\ldots,n\}}$ be an oriented link in an oriented rational homology sphere, then $\zeta(L)$ denotes the following coefficient:

• If $n = 1$, $\zeta(L) = \frac{1}{2} O_M(K_1)\Delta''(K_1)(1) - \dfrac{|H_1(M)|}{24\, O_M(K_1)^2}$

• If $n > 1$, $\zeta(L) = \dfrac{\partial^n}{\partial u_1 \partial u_2 \ldots \partial u_n} \mathcal{D}(L)(0, \ldots, 0) = \dfrac{\partial^n}{\partial t_1 \partial t_2 \ldots \partial t_n} \Delta(L)(1, \ldots, 1)$

According to the properties of the Alexander series (see §2.3), the ζ-coefficient of a link L of n components in an oriented rational homology sphere M satisfies:

1.4.2 $\zeta(L)$ depends neither on the orientation of the components of L nor on their order.

1.4.3 $\zeta(L\subset(-M)) = (-1)^{n-1}\,\zeta(L\subset M)$. ((-M) denotes the manifold M after an orientation reversal.)

1.4.4 If L is a split link, $\zeta(L)$ is zero.

The following definition can be skipped by the reader interested only in integral surgery presentations.

DEFINITION **1.4.5**: *Dedekind sums* (see [R-G])

Let p be an integer; let q be an integer or the (mod p)-congruence class of an integer, also denoted by q; the *Dedekind sum* s(q,p) is the following rational number:

$$s(q,p) = \sum_{i=1}^{|p|} \left(\left(\frac{i}{p}\right)\right)\left(\left(\frac{qi}{p}\right)\right) \text{ with } ((x)) = \begin{cases} 0 & \text{if } x \in \mathbb{Z}, \\ x - E(x) - \frac{1}{2} & \text{otherwise,} \end{cases}$$

where E(x) denotes the integer part of x.

First definition of $\mathbb{F}$ *(or first description of the surgery formula)*

(see 1.7.3 for an equivalent similar definition of $\mathbb{F}$ and 1.7.8 for a definition of $\mathbb{F}$ from one-variable polynomials)

1.4.6 *The 8-linking of* $\mathbb{L}_I$, denoted by $L_8(\mathbb{L}_I)$, is defined by:

$$L_8(\mathbb{L}_I) = \sum_{(j,\sigma)\in I\times\sigma_I} \left(\ell_{j\sigma(1)}\ell_{\sigma(1)\sigma(2)}\ell_{\sigma(2)\sigma(3)}\cdots\ell_{\sigma(i-1)\sigma(i)}\ell_{\sigma(i)j}\right)$$

where σ_I denotes the set of bijections from $\{1, \ldots, i = \#I\}$ to I.

Later on, the 8-linking of $\mathbb{L}_I$ will be seen as a sum running over all combinatorial ways of identifying the elements of I with the set of vertices of a graph G whose underlying space is the figure eight drawn in Figure 1.1. The summand corresponding to a graph G (whose vertices are labelled by I) is the linking of $\mathbb{L}$ with respect to this graph (see Definition 2.4.2).

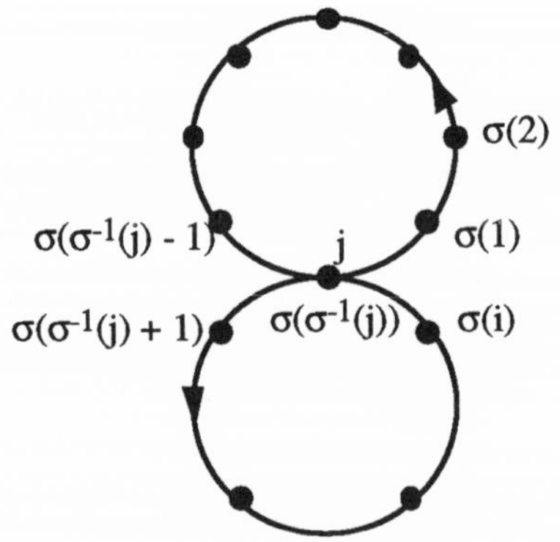

Figure 1.1

1.4.7 *The modified 8-linking of* $\mathbb{L}_I$, denoted by $M_8(\mathbb{L}_I)$, is defined by:

$$M_8(\mathbb{L}_I) = \begin{cases} L_8(\mathbb{L}_I) & \text{if } \#I > 1 \\ L_8(\mathbb{L}_{\{x\}}) + \dfrac{1}{q_x^2} = \dfrac{p_x^2 + 1}{q_x^2} & \text{if } I = \{x\} \end{cases}$$

1.4.8 *The function* $\mathbb{F}_M$ is defined on the set of all surgery presentations in a rational homology sphere M, with values in $\mathbb{Q}$, by:

$$\mathbb{F}_M(\mathbb{L}) =$$

$$\operatorname{sign}(\mathbb{L}) \left(\prod_{i=1}^{n} q_i \right) \sum_{\{ I \,/\, I \neq \emptyset \,,\, I \subset N \}} \det(E(\mathbb{L}_{N \setminus I})) \left(\zeta(L_I) + \frac{(-1)^{\#I} \, |H_1(M)| \, M_8(\mathbb{L}_I)}{24} \right)$$

$$+ \; |H_1(\chi(\mathbb{L}))| \left(\frac{\text{signature}\,(\,E(\mathbb{L})\,)}{8} + \sum_{i=1}^{n} \frac{s(p_i - q_i Lk_M(K_i,K_i), q_i)}{2} \right)$$

Recall that the determinant of an empty matrix equals one.

If $\mathbb{L}$ is an integral surgery presentation, the q_i's are one, thus the Dedekind sums $s(.,q_i)$ are zero.
When M is an integral homology sphere, the $Lk_M(K_i, K_i)$ are integers and can be ignored.

Recall also that $|H_1(\chi(\mathbb{L}))| = \operatorname{sign}(\mathbb{L}) \, |H_1(M)| \left(\prod_{i=1}^{n} q_i \right) \det(E(\mathbb{L}))$.

§1.5 Statement of the theorem

THEOREM

T1 There exists a function λ from the set of oriented closed 3-manifolds (up to orientation-preserving homeomorphisms) to $\frac{1}{12}\mathbb{Z}$, defined by:

For any surgery presentation $\mathbb{L}$ in S^3,

$$\lambda(\chi(\mathbb{L})) = \mathbb{F}_{S^3}(\mathbb{L})$$

The invariant λ satisfies the following properties:

T2 For any rational homology sphere M and for any surgery presentation $\mathbb{H}$ in M,

$$\lambda(\chi(\mathbb{H})) = \frac{|H_1(\chi(\mathbb{H}))|}{|H_1(M)|}\lambda(M) + \mathbb{F}_M(\mathbb{H})$$

T3 If (-M) denotes the manifold M equipped with the opposite orientation, and if $\beta_1(M)$ denotes the dimension of $H_1(M;\mathbb{Q})$:

$$\lambda(-M) = (-1)^{\beta_1(M)+1}\lambda(M)$$

T4 If # denotes the connected sum,

$$\lambda(M_1\#M_2) = |H_1(M_2)|\,\lambda(M_1) + |H_1(M_1)|\,\lambda(M_2)$$

T5 Let M be a closed oriented 3-manifold.

T5.0 If $\beta_1(M) = 0$, (i.e., if M is a rational homology sphere), and if λ_w denotes the Walker invariant as described in [W], then

$$\lambda(M) = \frac{|H_1(M)|}{2}\lambda_w(M)$$

(If M is an integral homology sphere and if λ_c denotes the Casson invariant as described in [G-M2], then

$$\lambda(M) = \lambda_c(M)\ )$$

T5.1 If $\beta_1(M) = 1$,

let $\Delta(M) \in \mathbb{Z}[t,t^{-1}]$ be the Alexander polynomial of M (i. e., the order of the first homology $\mathbb{Z}[t,t^{-1}]$-module of M with local coefficients in $\mathbb{Z}[t,t^{-1}] \cong \mathbb{Z}[\frac{H_1(M)}{\text{Torsion}}]$, normalized in order to be symmetric and to take a positive value at 1), then

$$\lambda(M) = \frac{\Delta''(M)(1)}{2} - \frac{|\text{Torsion}(H_1(M))|}{12}$$

T5.2 If $\beta_1(M) = 2$,

let S_1 and S_2 be two embedded surfaces, in general position in M, and such that their homology classes generate $H_2(M;\mathbb{Z})$; let γ be their oriented intersection, and let γ' be the curve parallel to γ inducing the trivialization of the tubular neighborhood of γ given by the surfaces; (note that γ and γ' are both rationally homologically trivial), then,

$$\lambda(M) = -\,|\mathrm{Torsion}(H_1(M))|\,Lk_M(\gamma,\gamma')$$

T5.3 If $\beta_1(M) = 3$,

Let {a, b, c} be a basis of $H^1(M;\mathbf{Z})$, let $\cup$ denote the cup product and let $[M]\in H_3(M;\mathbf{Z})$ be the orientation class of M, then

$$\lambda(M) = |\mathrm{Torsion}(H_1(M))|\left((a\cup b\cup c)([M])\right)^2$$

T5.≥4 If $\beta_1(M) \geq 4$,

$$\lambda(M) = 0$$

❑

§1.6 Sketch of the proof of the theorem and organization of the book

Preliminaries: Some essential facts about the Alexander series

We will begin in Chapter 2 with a definition of the Alexander series of an oriented link in an oriented rational homology sphere, derived from the homology of the maximal abelian covering of its exterior (or from the Reidemeister torsion of its exterior).

The following "change of variables property" will be clear from the definition: The Alexander series of two links with homeomorphic exteriors differ by an easily determined linear change of variables (and a possible multiplication by (-1)) (see 2.3.2).

1.6.1 If we call *"the truncated Alexander series"* of an n-component link its Alexander series up to degree n, this change of variables allows us to deduce the truncated Alexander series of a link from the truncated Alexander series of a link having the same exterior.

Alexander series satisfy restriction formulae (see 2.3.6) which relate the Alexander series of a link L to the Alexander series of its sublinks. For us, these restriction formulae have the two following important consequences.

1.6.2 Let L be a link with component indices in N. The series $\mathcal{D}(L_J)$, for $J\subset N$, can be reconstructed from $\mathcal{D}(L)$ and from the linking numbers $Lk(K_i,K_j)$, $\{i,j\}\subset N$, if all these linking numbers are nonzero; the same thing holds for the truncated Alexander series of these links.

1.6.3 The truncated Alexander series of a link L is a well-determined function of the linking numbers of the components of L and the ζ-coefficients of its sublinks (see 2.5.2).

The sketch of the proof

$\mathbb{F}_{S^3}$ will first be checked to define an invariant using the Rolfsen version (respectively the Fenn & Rourke version if we are only interested in integral presentations) of the Kirby calculus (see [F-R] and [Ro 2]). This version asserts that two surgery presentations in S^3 (considered up to ambient isotopy and changes in numbering and orientation of the components) present the same manifold if and only if one of them can be obtained from the other by a finite number of ω-twists (respectively integral ω-twists), and a finite number of their inverses, according to the following definition for the ω-twists.

DEFINITION **1.6.4**: *(ω-twist)*

Let $\mathbb{L} = (K_i , \frac{p_i}{q_i})_{i=0,\ldots,n}$ be a surgery presentation in a rational homology sphere M such that (a parallel of) the knot K_0 bounds an embedded disk D, parametrized by the unit disk of $\mathbb{C}$, in $M \setminus \overset{o}{T}(K_0)$, where $T(K_0)$ denotes a tubular neighborhood of K_0.

Let ω be an integer.

An *ω-twist with boundary* K_0 is the homeomorphism h_ω of $M \setminus \overset{o}{T}(K_0)$ which is the identity outside an oriented bicollar $D \times [-1,1]$ of D, and which maps $(z,t) \in D \times [-1,1]$ to $(e^{i\omega\pi(t+1)}z,t)$.

Such an ω-twist transforms $\mathbb{L}$ into the surgery presentation $\tilde{\mathbb{L}}$ in M described by:

$$\tilde{\mathbb{L}} = (\tilde{K}_i , (\tilde{p}_i / \tilde{q}_i))_{i=0,\ldots,n} \qquad \text{if } q_0 + \omega p_0 \neq 0,$$

$$\tilde{\mathbb{L}} = (\tilde{K}_i , (\tilde{p}_i / \tilde{q}_i))_{i=1,\ldots,n} \qquad \text{if } q_0 + \omega p_0 = 0,$$

where $\tilde{K}_i = h_\omega(K_i)$ for $i \in N = \{1, \ldots, n\}$, $\tilde{K}_0 = K_0$, and the characteristic curves are sent to their images under h_ω.

The surgery coefficients then become:

- $(\tilde{p}_i , \tilde{q}_i) = (p_i + Lk(K_i,K_0)^2\omega q_i , q_i)$ for $i \in \{1, \ldots, n\}$,
- $(\tilde{p}_0 , \tilde{q}_0) = (p_0 , q_0 + \omega p_0)\text{sign}(q_0 + \omega p_0)$ if $q_0 + \omega p_0 \neq 0$ (so that $\tilde{q}_0 > 0$).

(The orientation of μ_0 may change.)

It is clear from this definition that such an operation does not change the presented manifold.

An ω-twist is called an *integral ω-twist* if $\mathbb{L}$ is an integral surgery presentation and if $(q_0+\omega p_0) = 0$.

(Note that this definition includes deletions of trivial components with framing $-\frac{1}{\omega}$.)

Ambient isotopies, orientation reversals of the components, and permutations of the components of a surgery presentation $\mathbb{L}$ in S^3 do not affect $\mathbb{F}_{S^3}(\mathbb{L})$ (according to 1.4.2 and the symmetry in Definition 1.4.8 of $\mathbb{F}$).
To prove T1, it suffices then to prove that, under the assumptions and notation of Definition 1.6.4 of an ω-twist, Equality 1.6.5 holds:

$$\textbf{1.6.5} \qquad \mathbb{F}_{S^3}(\mathbb{L}) = \mathbb{F}_{S^3}(\tilde{\mathbb{L}})$$

Under the assumptions of T2, consider a surgery presentation $\mathbb{L}$ in S^3 for M and isotope $\mathbb{H}$ in M so that $\mathbb{H}$ is disjoint from the link of the cores of the surgery performed on $\mathbb{L}$. Let $\mathbb{H}_S$ denote the trace of $\mathbb{H}$ in S^3. As soon as T1 is proved, T2 is equivalent to the following equality 1.6.6:

$$\textbf{1.6.6} \qquad \mathbb{F}_{S^3}(\mathbb{H}_S \cup \mathbb{L}) = \frac{|H_1(\chi(\mathbb{H}))|}{|H_1(M)|}\, \mathbb{F}_{S^3}(\mathbb{L}) + \mathbb{F}_M(\mathbb{H})$$

Now, 1.6.5 and 1.6.6 are nothing but relations between ζ-coefficients of links with homeomorphic exteriors and ζ-coefficients of their restrictions. Hence they can be checked, at least generically (i.e. when some polynomials in the indeterminates $Lk(K_i,K_j)$ do not vanish), using only Consequences 1.6.1 to 1.6.3 of the "change of variables property" and of the restriction fomulae satisfied by the Alexander series.
1.6.5 and 1.6.6 will be proved in Chapters 3 and 4, respectively.

That $\lambda(M)$ belongs to $\frac{1}{12}\mathbb{Z}$ for any oriented closed 3-manifold M is proved in §6.3, where we also prove
6.3.3

$$\lambda(M) \in \frac{1}{4}\mathbb{Z} \Leftrightarrow H_1(M; \frac{\mathbb{Z}}{3\mathbb{Z}}) \neq \frac{\mathbb{Z}}{3\mathbb{Z}}$$

❑

About T3 *and* T4
Properties T3 and T4 can easily be obtained from the definition given by T1, using Properties 1.4.3 and 1.4.4 of the ζ-coefficients (respectively) together with appropriate surgery presentations.

About T5.0

Since any rational homology sphere can be obtained from S^3 by a finite sequence of integral surgeries on knots in such a way that every intermediate manifold is a rational homology sphere, it suffices to check that the Walker surgery formula and Formula T2 agree in the particular case of an integral surgery on a knot which yields a rational homology sphere to get T5.0. This is done in §4.7.

About T5.≥1

Any manifold M with positive first Betti number has a surgery presentation $\mathbb{H}$ in a rational homology sphere R, with a null linking matrix, and such that all the components of the underlying link H of $\mathbb{H}$ represent zero in $H_1(M)$. The surgery formula (T2) is in this case very simple and the computation of $\lambda(M)$ reduces to the computation of $\zeta(H)$ which is performed in Chapter 5 with Seifert surfaces techniques.

More about the organization of the book

§6.3 (re)proves

PROPOSITION 6.3.8: (Walker, Casson for the $\mathbb{Z}$-spheres)

The signature of a compact smooth spin 4-manifold with boundary a $\frac{\mathbb{Z}}{2\mathbb{Z}}$*- sphere* M *is congruent to* $8|H_1(M;\mathbb{Z})|\lambda(M)$ *mod* 16.

Next subsection (§1.7) states two equivalent definitions for $\mathbb{F}$:

Expression 1.7.3 of $\mathbb{F}$ looks less elegant than Expression 1.4.8, but it is more natural and it will be (more convenient to use and therefore) used until Chapter 6.

Expression 1.7.8 gives $\mathbb{F}$ as a function of linking numbers and coefficients of one-variable Alexander polynomials when the components of the surgery presentations are null-homologous.

The equivalence between all the given definitions of $\mathbb{F}$ is proved in Chapter 6.

The table of contents should now clearly explain what is in this book and where to find it.

Chapters 3, 4, 5, Sections §6.1, §6.2, §6.3 and §6.4 can be read independently. In particular, Chapter 4 can be read before Chapter 3, (assuming the invariance of $\mathbb{F}$ under a twist homeomorphism), it is easier and proves (even without the invariance of $\mathbb{F}$ under a twist homeomorphism) that the Walker invariant satisfies the surgery formula T2 for "generic" surgery presentations (see Remark 4.5.2).

§1.7 Equivalent definitions for $\mathbb{F}$

1.7.A Second equivalent definition of $\mathbb{F}$ (with relative ζ-coefficients)

DEFINITION **1.7.1:** (*Relative ζ-coefficients of* L)
Let J and I be two subsets of N, $J \subset I$, then $\zeta_I(L_J)$ is defined by:

$$\zeta_I(L_J) = \zeta(L_J) + \frac{(-1)^{\#J-1}|H_1(M)|}{24} \sum_{i \in I \setminus J} Lk_c(L_{J \cup \{i\}})$$

where $Lk_c(L_{J\cup\{i\}})$ is the *circular linking* of $L_{J\cup\{i\}}$ defined by:

1.7.2

$$Lk_c(L_{J\cup\{i\}}) = \sum_{\sigma \in \sigma_J} \ell_{i\sigma(1)}\ell_{\sigma(1)\sigma(2)}\ell_{\sigma(2)\sigma(3)}\cdots\ell_{\sigma(j-1)\sigma(j)}\ell_{\sigma(j)i}$$

Here σ_J denotes the set of bijections from $\{1, \ldots, j = \#J\}$ to J. (Compare with 2.4.9.)

1.7.3 *The function* $\mathbb{F}_M$ *can be defined from the set of the surgery presentations in the rational homology sphere* M *to* $\mathbb{Q}$ *by:*

$$\mathbb{F}_M(\mathbb{L}) =$$

$$\operatorname{sign}(\mathbb{L})\left(\prod_{i=1}^{n} q_i\right)\left(\sum_{\{I \,/\, I \neq \emptyset,\, I \subset N\}} \det(E(\mathbb{L}_{N \setminus I}))\,\zeta_N(L_I) - \sum_{i \in N} \det(E(\mathbb{L}_{N \setminus \{i\}}))\frac{|H_1(M)|}{24 q_i^2}\right)$$

$$+ |H_1(\chi(\mathbb{L}))|\left(\frac{\operatorname{signature}(E(\mathbb{L}))}{8} + \sum_{i=1}^{n} \frac{s(p_i - q_i Lk_M(K_i,K_i), q_i)}{2}\right)$$

$$- |H_1(\chi(\mathbb{L}))| \sum_{i=1}^{n} \frac{p_i}{24 q_i}$$

❑

The equivalence between 1.7.3 and 1.4.8 is proved in §6.2.

REMARK **1.7.4**:
Let ℓ_i be an oriented parallel of K_i, and let $<.,.>$ denote the intersection form on $\partial T(K_i)$, then

$$p_i - q_i Lk_M(K_i,K_i) \equiv <\mu_i, \ell_i> \qquad \text{mod } q_i$$

(This is proved in 6.2.B.)

1.7.B Third definition of $\mathbb{F}$ (in terms of one-variable Alexander polynomials)

NOTATION:

1.7.5 $$\Theta_b(\mathbb{L}_I) = \sum_{\{(K,i,j,g)\,/\,K \subset I,\ (i,j) \in K^2,\ g \in \sigma_{I\setminus K}\}} Lk_c(\mathbb{L}_K)\ell_{ig(1)}\ell_{g(1)g(2)}\cdots\ell_{g(\#(I\setminus K)-1)g(\#(I\setminus K))}\ell_{g(\#(I\setminus K))j}$$

Here $\sigma_{I\setminus K}$ denotes the set of bijections from $\{1, \ldots, \#(I\setminus K)\}$ to $I\setminus K$.

Roughly, we sum over all combinatorial ways of identifying the elements of I with the set of vertices of a graph G whose underlying space is one of the spaces drawn in Figure 1.2. The summand corresponding to a graph G (whose vertices are labelled by I) is the linking number of $\mathbb{L}$ with respect to this graph.

Figure 1.2 : Typical graphs involved in Θ_b

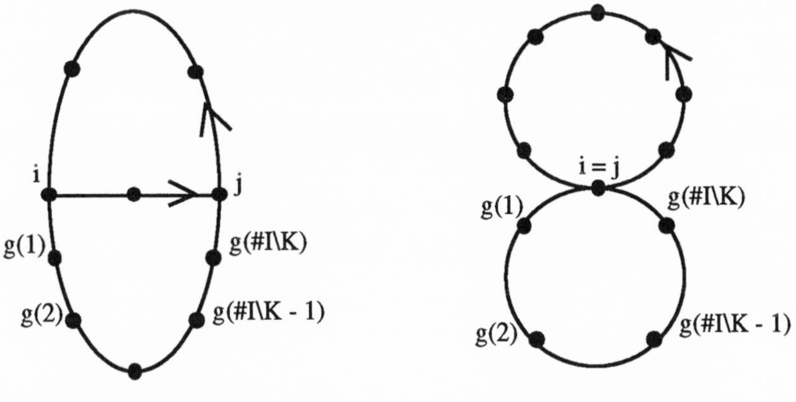

A beardless θ　　　　A figure-eight graph

1.7.6

$$\theta(\mathbb{L}_I) = \begin{cases} \Theta_b(\mathbb{L}_I) & \text{if } \#I > 2 \\ \Theta_b(\mathbb{L}_I) - 2\ell_{ij} & \text{if } I = \{i,j\} \\ \Theta_b(\mathbb{L}_I) + \dfrac{q_i^2 + 1}{q_i^2} & \text{if } I = \{i\} \end{cases}$$

1.7.7

$$E(\mathbb{L}_{N\setminus I}\,;\,I) = [\ell_{ijI}]_{i,j\in N\setminus I} \qquad \text{with } \ell_{ijI} = \begin{cases} \ell_{ij} & \text{if } i\neq j \\ \ell_{ii} + \sum_{k\in I} \ell_{ki} & \text{if } i=j \end{cases}$$

If L is a link of n null-homologous components, $a_1(L)$ is the coefficient

$$a_1(L) = \frac{\nabla_L^{(n+1)}(0)}{(n+1)!}$$

of its Conway polynomial ∇_L (see 2.3.15).

PROPOSITION **1.7.8**:
Let $\mathbb{L}$ *be a surgery presentation in a rational homology sphere* M *such that the components of the underlying link of* $\mathbb{L}$ *are null-homologous.*
Then, with the notation above, $\mathbb{F}_M(\mathbb{L})$ *can be written as*

$$\mathbb{F}_M(\mathbb{L}) =$$

$$\mathrm{sign}(\mathbb{L})\,|H_1(M)|\left(\prod_{i=1}^{n} q_i\right) \sum_{\{J\,/\,J\neq\emptyset\,,\,J\subset N\}} \det(E(\mathbb{L}_{N\setminus J};J))\; a_1(L_J)$$

$$+\,\mathrm{sign}(\mathbb{L})\,|H_1(M)|\left(\prod_{i=1}^{n} q_i\right) \sum_{\{J\,/\,J\neq\emptyset\,,\,J\subset N\}} \frac{\det(E(\mathbb{L}_{N\setminus J}))\;(-1)^{\#J}\;\theta(\mathbb{L}_J)}{24}$$

$$+\,|H_1(\chi_M(\mathbb{L}))|\left(\frac{\mathrm{signature}(E(\mathbb{L}))}{8} + \sum_{i=1}^{n} \frac{s(p_i - q_i Lk_M(K_i,K_i),q_i)}{2}\right)$$

❑

This will be proved in §6.4.

Chapter 2

The Alexander series of a link in a rational homology sphere and some of its properties

§2.1 The background

Alexander polynomials are classical invariants in knot theory and have been extensively studied.
The Alexander polynomial of a link in a rational homology sphere can be defined in the powerful and very appropriate context of Reidemeister torsion theory as a Reidemeister torsion of the exterior X of the link (up to a well-determined factor for a knot) and, following [Tu], it can be given a suitable sign, and hence a suitable normalization, if X is equipped with an orientation of $H_1(X;\mathbb{R})\oplus H_2(X;\mathbb{R})$.
The normalization of this Alexander polynomial for oriented links in S^3 was first pointed out by Conway and then established in a combinatorial way (by means of planar presentations) by Hartley (see [Hart]). Starting with Hartley's definition, Boyer and Lines gave a combinatorial (by means of surgery presentations of the ambient rational homology sphere) normalization of the Alexander polynomial of oriented links in any oriented rational homology sphere (see [B-L2]).

Section 2.2 gives a constructive definition of the normalized several-variable Alexander polynomial. This definition is only a way of computing the sign-determined torsion of [Tu] and coincides (up to an easily-determined factor) with the definition of Hartley when the ambient sphere is the usual one, and with the definition of Boyer and Lines in the general case. §2.3 states the general properties of the Alexander polynomial which are useful in this book, or which may help when applying the surgery formula or recognizing the normalization. Though §2.2 and §2.3 do not contain anything new, the results and assertions listed there will be proved in the appendix, for the sake of completeness, from the point of view of this book (suitable for explicit computations).

We will often work with Alexander series rather than Alexander polynomials. They represent the same invariant, but Alexander series are more convenient here to get linear changes of variables.

BRIDGE **2.1.1**: *(From Alexander series to Alexander polynomial and vice versa)*
The Alexander series $\mathcal{D}(L)$ of an oriented link $L = (K_1, \dots, K_n)$ in an oriented rational homology sphere M is a polynomial in $\mathbf{Z}[\exp(\pm\frac{u_i}{2O_M(K_i)})]_{i=1,\dots,n}$ (where exp is seen as the formal exponential series or as the Taylor series of the exponential map at 0).
It is, up to the change of variables,

$$\textbf{2.1.2} \qquad (t_i)^{1/2O_M(K_i)} = \exp(\frac{u_i}{2O_M(K_i)}) \qquad (t_i = \exp(u_i))$$

the several-variable normalized Alexander polynomial $\Delta(L)$:

$$\mathcal{D}(L)(u_1, u_2, \dots, u_n) = \Delta(L)(t_1, t_2, \dots, t_n)$$

(The change of variables 2.1.2 is either formal or local near $(u_i = 0)$ and $((t_i)^{1/2O_M(K_i)} = 1)$.)
In particular, if $n \geq 2$,

$$\zeta(L) = \frac{\partial^n \mathcal{D}(L)(0, \dots, 0)}{\partial u_1 \partial u_2 \dots \partial u_n} = \frac{\partial^n \Delta(L)(1, \dots, 1)}{\partial t_1 \partial t_2 \dots \partial t_n}$$

§2.5 contains the study of the Alexander series of an n-component link up to degree n. This study is the main tool of Chapters 3 and 4. §2.4 introduces the relevant functions of linking numbers.

§2.2 A definition of the Alexander series

Let $L = K_1, K_2, \dots, K_n$ be an oriented link of n components in an oriented rational homology sphere M. Let $m_1, m_2, \dots, m_n$ denote the oriented meridians (and their homology classes) of the components of L.
Then, any element x of $Q_{fa}(M\backslash L) = \frac{H_1(M\backslash L)}{\text{Torsion}(H_1(M\backslash L))}$ can be written as

$$x = \sum_{i=1}^{n} Lk_M(x,K_i)m_i \quad (Lk_M(x,K_i) \in \frac{1}{O_M(K_i)}\mathbf{Z})$$

This determines the ring morphism ψ_L:

2.2.1 $$\psi_L: \mathbf{Z}[\tfrac{1}{2}Q_{fa}(M\backslash L)] \to \mathbf{Z}[\exp(\pm\frac{u_i}{2O_M(K_i)})]_{i=1,\dots,n}$$
$$\exp(\tfrac{1}{2}x) \to \exp(\tfrac{1}{2}\sum_{i=1}^{n} Lk_M(x,K_i)u_i)$$

(See A.1.1 for the notation associated with group rings $\mathbf{Z}[.]$, recall that $\varepsilon: \mathbf{Z}[.] \to \mathbf{Z}$ denotes the *augmentation morphism* mapping exp(x) to 1.)

Let φ be the natural quotient map from $\pi_1(M\backslash L)$ to $Q_{fa}(M\backslash L)$.
If D is a CW-complex, which is a strong deformation retract of $M\backslash L$, let $C_*(D;\mathbf{Z}[Q_{fa}(M\backslash L)])$ denote the cellular complex associated with D, with local coefficients in $\mathbf{Z}[Q_{fa}(M\backslash L)]$, where $\mathbf{Z}[Q_{fa}(M\backslash L)]$ is equipped with its $\mathbf{Z}[\pi_1(M\backslash L)]$-module structure inherited from φ (see Remark 2.2.3).
Let D be a CW-complex which is a strong deformation retract of $M\backslash L$, the cells of which are:

- the basepoint of M, $*$,
- (k+1) one-cells, $c_1, c_2, \dots, c_{k+1}$, with $k \geq n-1$, such that (the closures of) $c_1, c_2, \dots, c_{n-1}$ are the oriented meridians of $K_1, K_2, \dots, K_{n-1}$ and the last cell c_{k+1} is the oriented meridian of K_n,
- k two-cells, $d_1, d_2, \dots, d_k$, such that (the closures of) $d_1, d_2, \dots, d_{n-1}$ represent the oriented boundaries of the tubular neighborhoods of $K_1, K_2, \dots, K_{n-1}$, respectively. (Such a D exists.)

Let $\partial(d_i) = \sum_{j=1}^{k+1} \partial_{ji}c_j$, with $\partial_{ji} \in \mathbf{Z}[Q_{fa}(M\backslash L)]$, denote the boundary of d_i in $C_*(D;\mathbf{Z}[Q_{fa}(M\backslash L)])$.
Then, there exists a positive unit

$$u \in \mathbf{Z}[\exp(\pm\frac{u_i}{2O_M(K_i)})]_{i=1,\dots,n}, \text{ (i.e. } u = \prod_{i=1}^{n}\exp(r_iu_i), \text{ with } r_i \in \frac{1}{2O_M(K_i)}\mathbf{Z})$$

such that the product

$$\mathcal{D}(L) = \text{sign}(\varepsilon(\det([\partial_{ij}]_{n\leq i,j\leq k})))\frac{u\,\psi_L\big(\det([\partial_{ij}]_{1\leq i,j\leq k})\big)}{(\exp(u_n)-1)}\,\delta_1(\exp(\frac{u_1}{O_M(K_1)})-1)$$

with $$\delta_1(\exp(\frac{u_1}{O_M(K_1)})-1) = \begin{cases} \exp(\frac{u_1}{O_M(K_1)}) - 1 & \text{if } n = 1 \\ 1 & \text{otherwise} \end{cases}$$

satisfies

$$\mathcal{D}(L)(u_1, u_2, \dots, u_n) = \pm\,\mathcal{D}(L)(-u_1, -u_2, \dots, -u_n)$$

DEFINITION **2.2.2**:
The element $\mathcal{D}(L)$ defined above is an invariant of the oriented link L in the oriented rational homology sphere M, called the *Alexander series of* L.

REMARK: The appendix (see §A.1 and §A.2) defines $\mathcal{D}(L)$ from any presentation of deficiency 1 of $H_1(M\setminus L,*;\mathbb{Z}[Q_{fa}(M\setminus L)])$ over $\mathbb{Z}[Q_{fa}(M\setminus L)]$. Definition 2.2.2 is only a particular case of this definition (resulting from A.1.6 and A.1.11). Nevertheless it always holds and it will be sufficient for our concerns.

REMARK **2.2.3**:
We can think of $C_*(D;\mathbb{Z}[Q_{fa}(M\setminus L)])$ as a cellular complex with the $\mathbb{Z}$-homology of the maximal free abelian covering of $M\setminus L$ (associated with φ).
In particular, $C_i(D;\mathbb{Z}[Q_{fa}(M\setminus L)])$ is freely generated over $\mathbb{Z}[Q_{fa}(M\setminus L)]$ by preferred liftings of the i-cells of D, which we will also denote like their projections; and the (equivariant under the action of $Q_{fa}(M\setminus L)$) boundary maps are computed in the covering mentioned above (see Example A.2.27).

PROPERTY **2.2.4**:
Each variable u_i of $\mathcal{D}(L)$ corresponds to a component K_i of L, but the set indexing the components does not have to be ordered. (Changing the order of the components of L does not affect $\mathcal{D}(L)$ in 2.2.1.)

EXAMPLE **2.2.5**: (see A.2.27)
Let $L = K_0, K_1, ..., K_n$ denote the following link in S^3 $(n\geq 1)$:

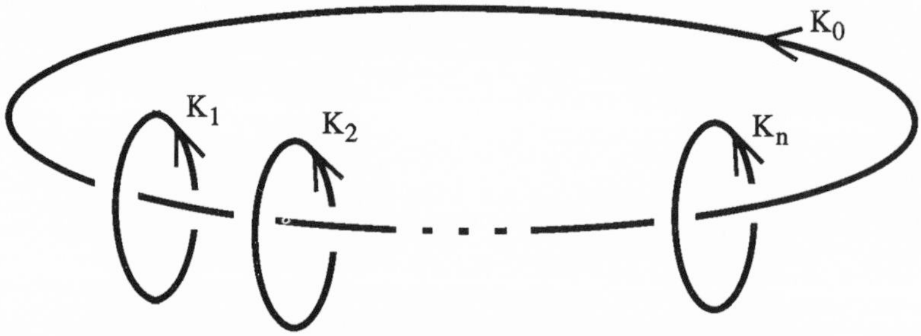

Figure 2.1

Then

$$\mathcal{D}(L) = (-1)^n \left(\exp(\frac{1}{2}u_0) - \exp(-\frac{1}{2}u_0)\right)^{n-1}$$

§2.3 A list of properties for the Alexander series

Let L be an oriented link of n components K_i, $i \in \{1, \ldots, n\}$, in an oriented rational homology sphere M.

Let $\mathcal{D}(L) \in \mathbb{Z}[\exp(\pm\frac{u_i}{2O_M(K_i)})]_{i=1,\ldots,n}$ be its Alexander series.

2.3.1 *If* $n=1$, $\mathcal{D}(L)$ *is independent of the orientations of* L *and* M *and it is symmetric.*
Furthermore,

$$\mathcal{D}(L)(0) = |\,\text{Torsion}\,(H_1(M\setminus L;\mathbb{Z}))\,|$$

From 2.3.2 to 2.3.11 we assume $n\geq 2$.

2.3.2 *Relation between Alexander series of two links with the same exterior*
Let L' be an oriented link in an oriented rational homology sphere M' such that $M\setminus L$ and $M'\setminus L'$ are (+)-homeomorphic. Let $\varepsilon = \pm 1$ be the sign of the change of basis from the basis of $H_1(M\setminus L;\mathbb{R})$ made of the oriented meridians of L to the basis of the oriented meridians of L' (with compatible orders). Then:

$$\mathcal{D}(L') = \varepsilon\, \psi_{L'}\circ(\psi_L)^{-1}\, \mathcal{D}(L)$$

(where ψ_L and $\psi_{L'}$ are as in 2.2.1).

Effect of various orientation changes
2.3.3 If L_{-1} denotes the link obtained from L by reversing the orientation of K_1, then:

$$\mathcal{D}(L_{-1})(u_1, u_2, \ldots, u_n) = -\,\mathcal{D}(L)(-u_1, u_2, \ldots, u_n)$$

2.3.4 If $\overline{L}$ denotes the link L in -M (i.e., when the orientation of the ambient space is reversed), then

$$\mathcal{D}(\overline{L})(u_1, u_2, \ldots, u_n) = -\,\mathcal{D}(L)(-u_1, -u_2, \ldots, -u_n)$$

2.3.5 *Symmetry*

$$\mathcal{D}(L)(u_1, u_2, \ldots, u_n) = (-1)^n \mathcal{D}(L)(-u_1, -u_2, \ldots, -u_n)$$

2.3.6 *Restriction formulae*
Let ℓ_{ij} denote the linking number of K_i and K_j in M.

$$\mathcal{D}(K_1, K_2)(u_1,0) = -\,\mathcal{D}(K_1)(u_1)\,\frac{\exp(\frac{\ell_{12}u_1}{2}) - \exp(-\frac{\ell_{12}u_1}{2})}{\exp(\frac{u_1}{2O_M(K_1)}) - \exp(-\frac{u_1}{2O_M(K_1)})}$$

If $n\geq 3$,

$$\mathcal{D}(L)(u_1, \ldots, u_{n-1}, 0)$$

$$= -\,\mathcal{D}(L_{N\setminus\{n\}})(u_1, \ldots, u_{n-1})\left(\exp(\sum_{i=1}^{n-1}\frac{\ell_{in}u_i}{2}) - \exp(-\sum_{i=1}^{n-1}\frac{\ell_{in}u_i}{2})\right)$$

2.3.7 *Nullity on split links*
If L is a split link, (that is, if there exists a sphere S^2, embedded in $M \setminus L$, which separates M into two components, each of which intersects L nontrivially), $\mathcal{D}(L)$ is zero.

2.3.8 *Transformation under addition of a meridian component*
If an oriented meridian K_0 of a component K_i is added to L, then

$$\mathcal{D}(L \cup K_0)(u_0, \ldots, u_n) = \left(\exp(-\frac{u_i}{2}) - \exp(\frac{u_i}{2})\right)\mathcal{D}(L)(u_1, \ldots, u_n)$$

(For further properties of this type see [Tu] Theorem 1.3.1, only 2.3.8 will be used in this book, for further properties of any type see also [Tu].)

2.3.9
If the components of L are null-homologous, then

$$\mathcal{D}(L) \in \prod_{i=1}^{n} \exp\left(\frac{u_i}{2}\left(\sum_{j \in \{1, \ldots, n\} \setminus \{i\}} \ell_{ij} + 1\right)\right) \mathbb{Z}[\exp(\pm u_i)]_{i=1, \ldots, n}$$

2.3.10 *Comparison with the Turaev refined Alexander function*
If $A_{\text{Turaev}}(L)$ denotes the Turaev refined Alexander function of L defined as in [Tu] §4.3, and next symmetrized, then

$$\Delta(L) = (-1)^n A_{\text{Turaev}}(L)$$

2.3.11 *Comparison with the Hartley and the Boyer and Lines normalizations*
If $M = S^3$,

$$\Delta(L) = (-1)^{n-1}\Delta_{\text{Hartley}}(L)$$

For any rational homology sphere M,

$$\Delta(L) = |H_1(M;\mathbb{Z})|(-1)^{n-1}\Delta_{\text{Boyer-Lines}}(L)$$

Relations with one-variable polynomials

DEFINITION **2.3.12:** Let L be an oriented null-homologous link in an oriented rational homology sphere M. The *Seifert form* of an oriented Seifert surface Σ of L is the bilinear form

$$\mathcal{V}: H_1(\Sigma;\mathbb{Z}) \times H_1(\Sigma;\mathbb{Z}) \to \mathbb{Q}$$

which assigns to any ordered pair ([x], [y]) of homology classes of two simple closed curves x and y in Σ, the linking number $Lk(x^+,y)$ in M, where x^+ denotes x pushed off the surface in the positive normal direction to Σ.

PROPOSITION **2.3.13**
Let L *be an oriented null-homologous n-component link in an oriented rational homology sphere* M, *and let* Σ *be an oriented connected Seifert surface of* L. *Let* V *be the matrix of the Seifert form of* Σ *with respect to a basis of* $H_1(\Sigma;\mathbb{Z})$. *Then:*

$$|H_1(M;\mathbb{Z})| \det(t^{1/2} V - t^{-1/2}\, {}^tV) = \begin{cases} \Delta(L)(t) & \text{if } n = 1 \\ \Delta(L)(t\,,\, t,\, \ldots,\, t)(t^{1/2} - t^{-1/2}) & \text{if } n \geq 2 \end{cases}$$

Furthermore, if $R_c(M\backslash L)$ *denotes the infinite cyclic covering of* $M\backslash L$ *associated with the "linking number with* L*",* $|H_1(M)|\det(V - t^{-1}\, {}^tV)$ *is the order of the* $\mathbb{Z}[t,t^{-1}]$*-module* $H_1(R_c(M\backslash L);\, \mathbb{Z})$.

CLAIM AND DEFINITION **2.3.14**: *The Conway polynomial and its coefficients*
With the notation of Proposition 2.3.13, $\det(t^{1/2} V - t^{-1/2}\, {}^tV)$ is a polynomial in $\mathbb{Q}[(t^{1/2} - t^{-1/2})]$, invariant of L, denoted by ∇_L, and called the Conway polynomial of L:

$$\nabla_L(z = t^{1/2} - t^{-1/2}) = \det(t^{1/2} V - t^{-1/2}\, {}^tV)$$

It can be written as in

2.3.15 $\quad \nabla_L(z) = z^{n-1}(a_0(L) + a_1(L)z^2 + \ldots + a_{d(L)}z^{2d(L)})$

This determines its coefficients $a_i(L)$.

2.3.16 *Skein relation for the Conway polynomial*
Let L^+, L^- and L^0 be three null-homologous links in M which are identical outside a ball of M where they look like:

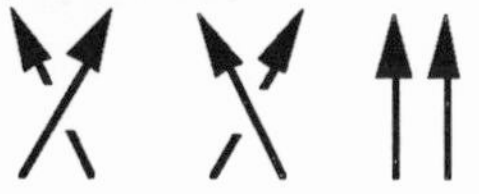

L^+ $\quad L^-$ $\quad L^0$

then, they satisfy the following skein relation:

$$\nabla_{L^+} - \nabla_{L^-} = - z\, \nabla_{L^0}$$

§2.4 Functions of the linking numbers of a link

The functions introduced in this section will be used in the following chapters.

Throughout this section, N denotes a finite set with cardinality n, and $L = (K_i)_{i\in N}$ denotes a link with component indices in N.

NOTATION **2.4.1**: (*for multiindices*)
A *multiindex* R with *support* a finite set S(R) = I is an element $(r_i)_{i\in I}$ of $\mathbb{N}^I$.

The *modulus* $|R|$ of such a multiindex R is defined by $|R|=\sum_{i\in I} r_i$.

$\mathbb{1}$ or $\mathbb{1}_N$ denotes the multiindex with support N whose coordinates are all equal to 1.

Diminution of a multiindex

Let R denote a multiindex $(r_i)_{i\in I}$, let $j \in I$ satisfy $r_j > 0$.
Then R_j will denote the multiindex $(r'_i)_{i\in I}$ defined by:

$$r'_i = \begin{cases} r_i & \text{if } i\neq j \\ r_i - 1 & \text{if } i=j \end{cases}.$$

We will denote $R_{ij} = (R_i)_j$, and so on, if it makes sense.

DEFINITION **2.4.2**: (*Linking of* L *with respect to a graph with support* N)
Let G be a graph with *support* N, that is, a graph with vertices $\{v_i\}_{i\in N}$.
Let E_{ij} denote an edge of G with ends v_i and v_j.
The linking of L *with respect to the edge* E_{ij} of G, denoted by $Lk(L;E_{ij})$ is defined by:

$$Lk(L;E_{ij}) = Lk(K_i,K_j)$$

If i=j, and if L is framed, $Lk(L;E_{ii})$ is the surgery coefficient $\frac{p_i}{q_i}$ of K_i.
The linking of L *with respect to the graph* G, denoted by Lk(L;G), is defined by:

$$Lk(L;G) = \prod_{\{\text{Edges } E \text{ of } G\}} Lk(L;E)$$

DEFINITION **2.4.3**: (*Linking of* L *with respect to a multiindex with support* N *and with modulus* (n-2))
Let R be a multiindex with support N and with modulus (n-2).
An R-*tree* is a tree with support N and such that, for $i\in N$, the *valence* of the vertex v_i is (r_i+1) (that is, v_i belongs to $(r_i + 1)$ edges). (R-trees exist if and only if $|R| = (n-2)$.)
The linking of L *with respect to* R, denoted by Lk(L;R), is defined by:

$$Lk(L;R) = \sum_{\{\text{R-trees } T\}} Lk(L;T)$$

EXAMPLES **2.4.4**:
Let $N = \{1, \ldots, n\}$, $R = (r_1, \ldots, r_n)$.
• If $r_1 = n-2$, and $r_i = 0$ $\forall i\geq 2$, then

$$Lk(L;R) = \prod_{i=2}^{n} Lk(K_1,K_i)$$

• If $r_1 = r_n = 0$, $r_i = 1$, $\forall i\in N\setminus\{1,n\}$ $(n\geq 2)$, and if $\mathfrak{B}$ is the set of cyclic permutations of N mapping n to 1, then

$$\mathrm{Lk}(L;R) = \sum_{b\in \mathcal{B}} \prod_{i=1}^{n-1} \mathrm{Lk}(K_i,K_{b(i)})$$

DEFINITION **2.4.5**: (*Linking of* L *with respect to a function)*
Let I and J be two subsets of N, and let f be a function from I to J.
Define the *graph* of f as the graph with support $I \cup J$ and with edges $\{E_{if(i)}\}_{i\in I}$.
The linking of L *with respect to* f, denoted by Lk(L;f), is the linking of L with respect to the graph of f, that is:

$$\mathrm{Lk}(L;f) = \prod_{i\in I} \mathrm{Lk}(K_i,K_{f(i)}).$$

DEFINITION AND NOTATION **2.4.6**: *(Attracted functions)*
Let I and J be two subsets of N.
A function f from I to $(I\cup J)$ is said to be *attracted* to J if:

For any element i of I, there exists a positive integer k such that $f^k(x)$ (is defined and) belongs to J.

Let $\mathcal{F}(I, \to J)$ denote the set of functions from I to $(I\cup J)$ which are attracted to J.

If R is a multiindex with support $I\cup J$, we say that such a function f *respects* (R,J) if:

For any element i of $I\cup J$, the cardinality of $f^{-1}(\{i\})$ is

$$\#f^{-1}(\{i\}) = \begin{cases} r_i & \text{if } i\notin J \\ r_i - 1 & \text{if } i\in J \end{cases}.$$

Let $\mathcal{F}(I, \to J;R)$ denote the set of functions from I to $(I\cup J)$ which are attracted to J and respect (R,J).
We will denote:

$$\mathrm{Lk}(L;I, \to J;R) = \sum_{f\in \mathcal{F}(I, \to J;R)} \mathrm{Lk}(L;f)$$

$$\mathrm{Lk}(L;I, \to J) = \sum_{f\in \mathcal{F}(I, \to J)} \mathrm{Lk}(L;f)$$

REMARK **2.4.7**: *About the empty set*
An empty product equals one (an empty sum equals zero), and there is one function from the empty set to any set. So, for example the determinant of an empty matrix equals one, being the sum over the permutations of the empty set of the (empty) products corresponding to these permutations; and, as another example

$$\mathrm{Lk}(L;\emptyset, \to N; \mathbb{1}) = 1$$

EXAMPLES **2.4.8**:

1. Let i be an element of N, let R be a multiindex with support N, with modulus n and with $r_i \geq 2$. Then:

$$Lk(L;N\setminus\{i\}, \to\{i\};R) = Lk(L;R_{ii})$$

2. If there exists $j \in J$, such that r_j is zero, or if I is not empty and if r_j equals 1 for all $j \in J$, then:

$$Lk(L;I, \to J;R) = 0$$

DEFINITION **2.4.9**: (*Circular linking of* L)

An *oriented circle* with support N is a graph with support N which is topologically a circle, equipped with an orientation.

The circular linking of L, denoted by $Lk_c(L)$, is defined by:

$$Lk_c(L) = \sum_{\{\text{Oriented circles } \mathfrak{C} \text{ with support } N\}} Lk(L;\mathfrak{C})$$

That is, if $\mathfrak{C}_N$ denotes the set of cyclic permutations of N,

$$Lk_c(L) = \sum_{\sigma \in \mathfrak{C}_N} Lk(L;\sigma)$$

§2.5 The first terms of the Alexander series

NOTATION **2.5.1**:

If $R = (r_1, r_2, ..., r_n)$, and if $u = (u_1, u_2, ..., u_n)$, u^R denotes the product:

$$u^R = \prod_{i=1}^{n} u_i^{r_i}$$

and $\mathfrak{D}_R(L)$ denotes the coefficient of u^R in $\mathfrak{D}(L)(u) = \sum_{R \in \mathbb{N}^n} \mathfrak{D}_R(L)\, u^R$.

Note that, if $n \geq 2$,

$$\mathfrak{D}_{\mathbb{1}}(L) = \frac{\partial^n}{\partial u_1 \partial u_2 \ldots \partial u_n} \mathfrak{D}(L)(0, \ldots, 0) = \zeta(L)$$

The following proposition expresses the truncated Alexander series of a link L of $n \geq 2$ components (up to order n) as a function of its linking numbers and the coefficients ζ of its sublinks.

Recall that

$$\zeta_I(L_J) = \zeta(L_J) + \frac{(-1)^{\#J-1}|H_1(M)|}{24} \sum_{i \in I\setminus J} Lk_c(L_{J \cup \{i\}})$$

and that

$$\zeta(K_j) = \frac{1}{2} O_M(K_j) \mathcal{D}''(K_j)(0) - \frac{|H_1(M)|}{24\, O_M(K_j)^2}$$

PROPOSITION **2.5.2**:
Let L *be a link of* $n \geq 2$ *components* K_i, $i \in N = \{1, \ldots, n\}$ *in a rational homology sphere* M.
Let $R = (r_1, \ldots, r_n)$ *be a multiindex with support* N.

- If $|R| < n - 2$, $\mathcal{D}_R(L) = 0$
- If $|R| = n - 2$, $\mathcal{D}_R(L) = (-1)^{n-1} |H_1(M)| Lk(L;R)$
- If $|R| = n - 1$, $\mathcal{D}_R(L) = 0$
- If $|R| = n$, $\mathcal{D}_R(L) = \sum_{\{I \,/\, I \subset N\}} \left((-1)^{n - \#I} Lk(L; N\backslash I, \to I; R)\, \zeta_N(L_I) \right)$

PROOF: Observe first that the proposition is true if $R = \mathbb{1}$.
Proceed then by induction on the number n of components of the link.
At each induction step, because of the first remark and the symmetry of the formulae, it suffices to prove the proposition for multiindices R such that $r_n = 0$.
To do this, we use the restriction formulae 2.3.6. We set:

$$Lk(K_i, K_j) = \ell_{ij}$$

The case (n=2)
If n=2, the restriction formula yields:
$\mathcal{D}(L)(u,0)$

$$= - \frac{u\,\ell_{12} + \dfrac{(u\,\ell_{12})^3}{24} + O(u^5)}{\dfrac{u}{O_M(K_1)} + \dfrac{u^3}{24\, O_M(K_1)^3} + O(u^5)} \left(\frac{|H_1(M)|}{O_M(K_1)} + \frac{\Delta''(K_1)(1)}{2} u^2 + O(u^4) \right)$$

$$= - |H_1(M)|\, \ell_{12}$$

$$+ \left(\frac{\ell_{12} |H_1(M)|}{24\, O_M(K_1)^2} - \frac{\ell_{12}\, O_M(K_1) \Delta''(K_1)(1)}{2} - \frac{(\ell_{12})^3 |H_1(M)|}{24} \right) u^2 + O(u^4)$$

$$= - |H_1(M)| Lk(L;(0,0)) - Lk(L; \{2\}, \to\{1\}; (2,0))\, \zeta_{\{1,2\}}(L_{\{1\}})\, u^2 + O(u^4)$$

The proposition is true for n=2.

The induction step
Assume now that $n \geq 3$ and that the proposition is true for $L_{\{1,\ldots,n-1\}}$.

The restriction formula 2.3.6 leads to

2.5.3 $$\mathcal{D}(L)(u_1, u_2, \ldots, u_{n-1}, 0) = -\left(\sum_{i=1}^{n-1} \ell_{in} u_i + \frac{(\sum_{i=1}^{n-1} \ell_{in} u_i)^3}{24} + O(5)\right) \mathcal{D}(L_{\{1,\ldots,n-1\}})(u_1, u_2, \ldots, u_{n-1})$$

The cases when $|R| < (n-2)$ and $|R| = n-1$ follow easily from 2.5.3.

• If $|R| = (n-2)$ and $r_n = 0$.
According to 2.5.3,

$$\mathcal{D}_R(L) = -\sum_{i=1}^{n-1} \ell_{in}\, \mathcal{D}_{R_i}(L_{N\setminus\{n\}})$$

(where the natural restriction of R_i to a multiindex with support $N\setminus\{n\}$ is still denoted by R_i, and, if R_i is not a mutiindex anymore (that is, if r_i=0), the term involving R_i is set to be zero).
Now, since each R-tree is obtained by joining n, by an edge E_{in}, to an R_i-tree with support $N\setminus\{n\}$,

$$Lk(L;R) = \sum_{i=1}^{n-1} \ell_{in}\, Lk(L_{N\setminus\{n\}};R_i)$$

This proves the proposition for $|R| = n-2$.

• If $|R| = n$ and $r_n = 0$, by 2.5.3 we have:

$$\mathcal{D}_R(L) = -\sum_{i=1}^{n-1} \ell_{in}\, \mathcal{D}_{R_i}(L_{N\setminus\{n\}}) - \frac{1}{24} \sum_{(i,j,k)\in(N\setminus\{n\})^3} \ell_{in}\ell_{jn}\ell_{kn}\, \mathcal{D}_{R_{ijk}}(L_{N\setminus\{n\}})$$

(where $\mathcal{D}_{R_{ijk}}(L_{N\setminus\{n\}})$ is set to be zero if R_{ijk} is not a multiindex).
Now, note that:

$$Lk(L;N\setminus I,\to I;R) = \begin{cases} 0 & \text{if } n \in I, \\ \sum_{i=1}^{n-1} \ell_{in}\, Lk(L_{N\setminus\{n\}};(N\setminus\{n\})\setminus I,\to I;R_i) & \text{otherwise.} \end{cases}$$

So, according to the induction hypothesis:

$$\sum_{\{I\,/\,I\subset N\}} \Big((-1)^{n-\#I}\, Lk(L;\, N\setminus I, \to I;R)\, \zeta_{N\setminus\{n\}}(L_I)\Big) = -\sum_{i=1}^{n-1} \ell_{in}\, \mathcal{D}_{R_i}(L_{N\setminus\{n\}}).$$

We are thus left with the proof of:

$$\sum_{\{I\,/\,I\subset N\}} \Big((-1)^{n-\#I}\, Lk(L;\, N\setminus I, \to I;R)\, (\, \zeta_N(L_I) - \zeta_{N\setminus\{n\}}(L_I)\,)\Big)$$
$$= -\frac{1}{24} \sum_{(i,j,k)\in(N\setminus\{n\})^3} \ell_{in}\ell_{jn}\ell_{kn}\, \mathcal{D}_{R_{ijk}}(L_{N\setminus\{n\}}),$$

which is equivalent to 2.5.4:

2.5.4
$$\sum_{\{I\, /\, I\subset N\setminus\{n\}\}} \Big(Lk(L;\, N\setminus I,\, \to I;R)\ \big(Lk_c(L_{I\cup\{n\}})\big) \Big) = \sum_{(i,j,k)\in (N\setminus\{n\})^3} \ell_{in}\ell_{jn}\ell_{kn}\, Lk(L_{N\setminus\{n\}};R_{ijk})$$

To prove 2.5.4, a correspondence between the terms of the two sides of the equality will be established.
The left-hand side is the sum running over all triples (I, g, γ) such that:

- I is a non-empty subset of $N\setminus\{n\}$,
- $g\in \mathfrak{F}(N\setminus I, \to I;R)$,
- γ is an oriented circle with support $I\cup\{n\}$

of the terms $Lk(L;g)Lk(L;\gamma)$.
The right-hand side is the sum running over all pairs ((i,j,k),T) such that:

- (i,j,k) is a triple of $(N\setminus\{n\})^3$ such that R_{ijk} is still a multiindex and,
- T is an R_{ijk}-tree with support $N\setminus\{n\}$,

of the terms $\ell_{in}\ell_{jn}\ell_{kn}\, Lk(L_{N\setminus\{n\}};T)$.

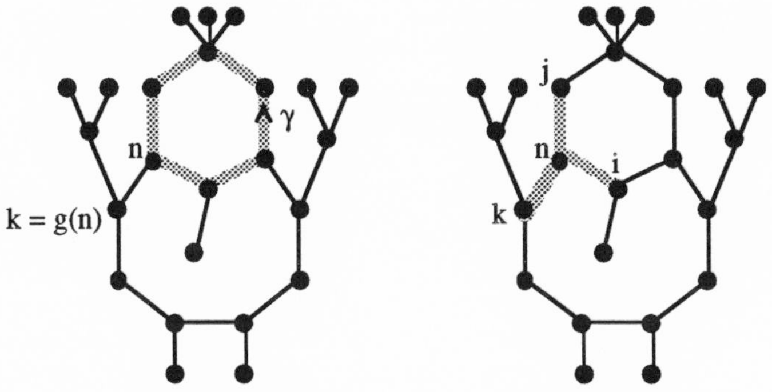

Grey γ and black graph of g Black T

Figure 2.2

Associate with a pair ((i,j,k),T) as above, the triple (I, g, γ) defined as follows:
Let [i,j] denote the oriented segment from i to j in T. Define I=I((i,j,k),T) as its support.
Define $\gamma=\gamma((i,j,k),T)$ as the oriented (by [i,j] if $i\neq j$) circle $[i,j]\cup E_{in}\cup E_{jn}$. The support of γ is $I\cup\{n\}$.
Let $T'= E_{kn} \cup T$ be the R_{ij}-tree with support N obtained by adding an edge E_{kn} to T.

Removing the edges of [i,j] from T' transforms T' into the graph of a unique function g((i,j,k),T) of $\mathfrak{F}$(N\I, →I;R). This defines g=g((i,j,k),T).

The triple (I, g, γ) satisfies:

$$Lk(L;g)Lk(L;\gamma) = \ell_{in}\ell_{jn}\ell_{kn}\, Lk(L_{N\setminus\{n\}};T)$$

Furthermore, the correspondence defined from the terms of the right-hand side to the terms of the left-hand side is bijective. This proves 2.5.4 and hence the proposition.

❑

REMARK **2.5.5**: Knowing the Alexander series $\mathcal{D}$(L) of L up to degree d is equivalent to knowing the Taylor series of Δ(L) at (1, ..., 1) up to order d. In particular, it is easy to recover the Taylor series of the Alexander polynomial of an n-component link L up to order n from Proposition 2.5.2.

Chapter 3

Invariance of the surgery formula under a twist homeomorphism

§3.1 Introduction

This chapter is entirely devoted to the proof of Proposition 3.1.1, that is the invariance of $\mathbb{F}_M$ under the ω-twist described in Definition 1.6.4.

PROPOSITION **3.1.1**: *(Invariance of* $\mathbb{F}$ *under an ω-twist)*
With the notation of Definition 1.6.4,

$$\mathbb{F}_M(\mathbb{L}) = \mathbb{F}_M(\tilde{\mathbb{L}})$$

❑

As we observed in the remarks before Equality 1.6.5, the invariance of $\mathbb{F}_{S^3}$ under these twist homeomorphisms is sufficient to ensure that $\mathbb{F}_{S^3}$ defines an invariant of closed oriented 3-manifolds. Proving the invariance of $\mathbb{F}_M$ for any rational homology sphere M is not more difficult, and it will allow us to consider only generic presentations when proving the general surgery formula T2 in Chapter 4.

The chapter is organized as follows:
§3.2 decomposes $\mathbb{F}_M(\mathbb{L})$ into pieces and describes the variation of the different pieces under the ω-twist. After §3.3, the proof will be reduced to the study of the variation of the piece containing the ζ-coefficients. The study of this variation is performed in §3.6. It uses new linking functions introduced in §3.4, and a description of the variation of the ζ-coefficients under an ω-twist performed in §3.5.

Throughout this chapter, we will use all the notation of Definition 1.6.4 as well as:
NOTATION **3.1.2**:
• N denotes the set of integers from 1 to n.

• ℓ_{ij} (respectively $\tilde{\ell}_{ij}$) denotes the linking number of K_i and K_j (respectively of $\tilde{K}_i$ and $\tilde{K}_j$) if i and j are different, and ℓ_{ii} (respectively $\tilde{\ell}_{ii}$) denotes the coefficient $\frac{p_i}{q_i}$ (respectively $\frac{\tilde{p}_i}{\tilde{q}_i}$).

So, the relation

$$\textbf{3.1.3} \qquad \tilde{\ell}_{ij} = \ell_{ij} + \omega \ell_{i0}\ell_{j0}$$

holds for any positive i and j.

To see 3.1.3, it suffices to write the characteristic curve μ_j as a combination of the meridians m_0 and m_i of K_0 and K_i, in $H_1(M\setminus(K_0 \cup K_i);\mathbb{Q})$

$$\frac{1}{q_j}\mu_j = \ell_{ij}\, m_i + \ell_{j0}\, m_0$$

So, in $H_1(M\setminus(K_0 \cup \tilde{K}_i);\mathbb{Q})$

$$\begin{aligned}\frac{1}{q_j} h_\omega(\mu_j) &= \ell_{ij}\, h_\omega(m_i) + \ell_{j0}\, h_\omega(m_0) = \ell_{ij}\, \tilde{m}_i + \ell_{j0}\, (m_0 + \omega\ell_{i0}\, \tilde{m}_i) \\ &= (\ell_{ij} + \omega\ell_{i0}\ell_{j0})\, \tilde{m}_i + \ell_{j0} m_0\end{aligned}$$

❑

§3.2 Variation of the different pieces of $\mathbb{F}_M$ under an ω-twist: the statements

About the case $q_0 + \omega p_0 = 0$

3.2.1 Recall that, in this case, K_0 does not appear in $\tilde{\mathbb{L}}$, according to Definition 1.6.4:

$$\tilde{\mathbb{L}} = \tilde{\mathbb{L}}_N$$

Nevertheless, we still give a meaning to $\tilde{L}_{I\cup\{0\}}$ by setting $\tilde{L}_{\{0\}}=K_0$ (as when $q_0+\omega p_0 \neq 0$). We also make natural sense of the algebraic expressions involving the framing of $\tilde{K}_0$, which we continue to formally denote by $p_0/(q_0+\omega p_0)$, as soon as the result of their formal calculation does not contain $q_0+\omega p_0$ in its denominator.

For example, if $q_0+p_0\omega = 0$,

$$\textbf{3.2.2} \qquad (q_0+p_0\omega)\det(E(\tilde{\mathbb{L}}_{N\cup\{0\}})) = p_0\det(E(\tilde{\mathbb{L}}_N))$$

So, **in any case**, if I is a subset of N, we have

$$\textbf{3.2.3} \qquad (q_0+p_0\omega)\det(E(\tilde{\mathbb{L}}_{I\cup\{0\}})) \quad = \quad q_0\det(E(\mathbb{L}_{I\cup\{0\}}))$$

3.2.4 PROOF OF 3.2.3:

Call $F_0(\tilde{\mathbb{L}})$ (respectively $F_0(\mathbb{L})$) the matrix obtained from $E(\tilde{\mathbb{L}}_{N\cup\{0\}})$ (respectively from $E(\mathbb{L})$) by multiplying the 0th column of $E(\tilde{\mathbb{L}}_{N\cup\{0\}})$ by $q_0+\omega p_0$ (respectively of $E(\mathbb{L})$ by q_0).

Then, $F_0(\tilde{\mathbb{L}})$ (which is defined even if $q_0+\omega p_0 = 0$) is obtained from $F_0(\mathbb{L})$ by adding, for $i\in N$, $\omega\ell_{i0}$ times the 0th row of $F_0(\mathbb{L})$ to the ith row of $F_0(\mathbb{L})$. So, $F_0(\tilde{\mathbb{L}})$ and $F_0(\mathbb{L})$ have the same determinant and this proves 3.2.3.

❑

LEMMA **3.2.5**

$$\mathrm{sign}(\tilde{\mathbb{L}}) = \begin{cases} \mathrm{sign}(\mathbb{L})\mathrm{sign}(q_0+p_0\omega) & \text{if } q_0+p_0\omega \neq 0 \\ p_0\mathrm{sign}(\mathbb{L}) & \text{if } q_0+p_0\omega = 0 \end{cases}$$

PROOF:

(a) If $\det(E(\mathbb{L}))$ is nonzero, $\mathrm{sign}(\mathbb{L}) = \mathrm{sign}(\det(E(\mathbb{L})))$, so 3.2.2 and 3.2.3 prove Lemma 3.2.5.

(b) If $p_0 = 0$, then $E(\mathbb{L})$ and $E(\tilde{\mathbb{L}})$ represent the same bilinear form, and the lemma is true.

(c) If none of the cases above holds, there exists a set $I\subset N$, such that $E(\mathbb{L}_{I\cup\{0\}})$ is non-degenerate and has the same rank $(=\#I+1)$ as $E(\mathbb{L})$. So,

$$\mathrm{sign}(\mathbb{L}) = \mathrm{sign}(\mathbb{L}_{I\cup\{0\}})$$

and,

$$\mathrm{sign}(\tilde{\mathbb{L}}) = \mathrm{sign}(\tilde{\mathbb{L}}_{I\cup\{0\}}) \qquad (\text{or } \mathrm{sign}(\tilde{\mathbb{L}}) = \mathrm{sign}(\tilde{\mathbb{L}}_I) \text{ if } q_0+p_0\omega = 0)$$

and, according to (a), the lemma is true if $\mathbb{L}$ is replaced by $\mathbb{L}_{I\cup\{0\}}$.

❑

Let us point out that

$$\textbf{3.2.6} \qquad p_0^2 = 1 \qquad \text{if } q_0+p_0\omega = 0$$

So, the following is a direct consequence of 3.2.2 and 3.2.5:

CONSEQUENCE **3.2.7**: *For any subset* I *of* N,

$$\mathrm{sign}(\mathbb{L})(q_0+p_0\omega)\det(E(\tilde{\mathbb{L}}_{I\cup\{0\}}))$$
$$= \begin{cases} \mathrm{sign}(\tilde{\mathbb{L}})|q_0+p_0\omega|\det(E(\tilde{\mathbb{L}}_{I\cup\{0\}})) & \text{if } q_0+p_0\omega \neq 0 \\ \mathrm{sign}(\tilde{\mathbb{L}})\det(E(\tilde{\mathbb{L}}_I)) & \text{if } q_0+p_0\omega = 0 \end{cases}$$

❑

Now, we can use Definition 1.7.3 and Remark 1.7.4 to write $\mathbf{F}_M(\tilde{\mathbb{L}})$ as in

LEMMA **3.2.8**: *Decomposition of* $\mathbb{F}_M(\tilde{\mathbb{L}})$

$$\mathbb{F}_M(\tilde{\mathbb{L}}) = \operatorname{sign}(\mathbb{L})\left(\prod_{i=1}^{n} q_i\right)(a+b)(\tilde{\mathbb{L}}) + |H_1(\chi_M(\mathbb{L}))|(c+d+e)(\tilde{\mathbb{L}})$$
$$+ \operatorname{sign}(\mathbb{L})\left(\prod_{i=1}^{n} q_i\right)|H_1(M)|f(\tilde{\mathbb{L}})$$

where

$$a(\tilde{\mathbb{L}}) = \sum_{\{I\,/\,I\neq\emptyset\,,\,I\subset N\cup\{0\}\}} (q_0+p_0\omega)\det(E(\tilde{\mathbb{L}}_{N\cup\{0\}\setminus I}))\zeta_{N\cup\{0\}}(\tilde{L}_I)$$

$$b(\tilde{\mathbb{L}}) = -|H_1(M)|(q_0+p_0\omega)\sum_{i=1}^{n}\frac{\det(E(\tilde{\mathbb{L}}_{N\cup\{0\}\setminus\{i\}}))}{24q_i^2}$$

$$c(\tilde{\mathbb{L}}) = \frac{\operatorname{signature}(E(\tilde{\mathbb{L}})) - \operatorname{sign}(p_0(q_0+p_0\omega))}{8}$$

$$d(\tilde{\mathbb{L}}) = \frac{1}{2}\sum_{i=1}^{n} s(<\mu_i,\ell_i>,q_i)$$

$$e(\tilde{\mathbb{L}}) = -\sum_{i=1}^{n}\frac{\tilde{p}_i}{24q_i}$$

and f *contains the contributions of* K_0 *(omitted until now), that is:*
If $q_0+p_0\omega \neq 0$,

$$f(\tilde{\mathbb{L}}) = -\frac{\det(E(\tilde{\mathbb{L}}_N))}{24(q_0+p_0\omega)}$$
$$+\frac{q_0\det(E(\mathbb{L}))}{2}\left(\frac{\operatorname{sign}(p_0(q_0+p_0\omega))}{4}+\frac{s(p_0,q_0+p_0\omega)}{\operatorname{sign}(q_0+p_0\omega)}-\frac{p_0}{12(q_0+p_0\omega)}\right)$$

If $q_0+p_0\omega = 0$,

$$f(\tilde{\mathbb{L}}) = -\frac{1}{24p_0}\sum_{\{I\,|\,I\subset N,\,I\neq\emptyset\}}\det(E(\tilde{\mathbb{L}}_{N\setminus I}))(-1)^{\#I-1}Lk_c(\tilde{L};I\cup\{0\})$$

PROOF: It suffices to copy Definition 1.7.3, and to use 1.3.5, 3.2.2, 3.2.7, and:

$$\chi_M(\tilde{\mathbb{L}}) = \chi_M(\mathbb{L})$$

Note that, when $q_0+p_0\omega = 0$, $\tilde{\mathbb{L}} = \tilde{\mathbb{L}}_N$. So, the relative ζ-coefficients involved in $\mathbb{F}_M(\tilde{\mathbb{L}})$ are the ζ_N. In this case, $f(\tilde{\mathbb{L}})$ makes up for the use of the $\zeta_{N\cup\{0\}}$ instead of the ζ_N in 3.2.8.

❑

NOTATION **3.2.9**: (*For a similar decomposition of* $\mathbb{F}_M(\mathbb{L})$)

For g = a, b, ..., f, we define $g(\mathbb{L})$ by removing the ~ and replacing $(q_0+p_0\omega)$ by q_0 in the definition of $g(\tilde{\mathbb{L}})$ introduced in 3.2.8. So, for example

$$f(\mathbb{L}) = \frac{q_0}{2}\det(E(\mathbb{L}))\left(\frac{\operatorname{sign}(p_0q_0)}{4} + s(p_0,q_0) - \frac{p_0}{12q_0}\right) - \frac{\det(E(\mathbb{L}_N))}{24q_0}$$

We can now write $\mathbb{F}_M(\mathbb{L})$ under a form similar to the form of $\mathbb{F}_M(\tilde{\mathbb{L}})$ given in 3.2.8.

3.2.10
$$\mathbb{F}_M(\mathbb{L}) = \operatorname{sign}(\mathbb{L})\left(\prod_{i=1}^{n} q_i\right)(a+b)(\mathbb{L}) + |H_1(\chi_M(\mathbb{L}))|(c+d+e)(\mathbb{L})$$
$$+ \operatorname{sign}(\mathbb{L})\left(\prod_{i=1}^{n} q_i\right)|H_1(M)|f(\mathbb{L})$$

❑

Now, according to the decompositions 3.2.8 and 3.2.10 of $\mathbb{F}_M(\tilde{\mathbb{L}})$ and $\mathbb{F}_M(\mathbb{L})$, Proposition 3.1.1 will be the consequence of Equalities 3.2.11 to 3.2.16:

3.2.11 $$a(\tilde{\mathbb{L}}) - a(\mathbb{L}) = \frac{|H_1(M)|\omega q_0 \det(E(\mathbb{L}_{N\cup\{0\}}))}{24}\left(\sum_{i=1}^{n}\ell_{i0}^2 - 1\right)$$

3.2.12 $b(\tilde{\mathbb{L}}) = b(\mathbb{L})$

3.2.13 $|H_1(\chi_M(\mathbb{L}))|\, c(\tilde{\mathbb{L}}) = |H_1(\chi_M(\mathbb{L}))|\, c(\mathbb{L})$

3.2.14 $d(\tilde{\mathbb{L}}) = d(\mathbb{L})$

3.2.15 $$e(\tilde{\mathbb{L}}) - e(\mathbb{L}) = -\frac{\omega}{24}\sum_{i=1}^{n}\ell_{i0}^2$$

3.2.16 $$f(\tilde{\mathbb{L}}) - f(\mathbb{L}) = \frac{\omega q_0 \det(E(\mathbb{L}))}{24}$$

❑

Equality 3.2.15 follows from 3.1.3.
Equality 3.2.12 follows from 3.2.3 and the invariance of the q_i's.
Equality 3.2.14 is obvious (see Remark 1.7.4).

❑

Proposition 3.1.1 is now proved except for Equalities 3.2.13 and 3.2.16 which are proved in §3.3 and Equality 3.2.11 which is proved in §3.6.

❑

§3.3 Proofs of 3.2.13 and 3.2.16

NOTATION **3.3.1**:

E_0 (respectively $\tilde{E}_0$) denotes the matrix obtained from $E(\mathbb{L})$ (respectively from $E(\tilde{\mathbb{L}})$) by replacing ℓ_{00} (respectively $\tilde{\ell}_{00}$) by 0 and keeping the other coefficients unchanged. As in 3.2.4, $\det(E_0) = \det(\tilde{E}_0)$.

PROOF OF 3.2.16

The following formal equalities between polynomials in the indeterminates p_0, ω, q_0 and the ℓ_{ij} $(i,j)\neq(0,0)$:

$$q_0\det(E_0) + p_0\det(E(\mathbb{L}_N)) = q_0\det(E(\mathbb{L})) = (q_0+p_0\omega)\det(E(\tilde{\mathbb{L}}_{N\cup\{0\}}))$$
$$= (q_0+p_0\omega)\det(E_0) + p_0\det(E(\tilde{\mathbb{L}}_N))$$

lead to the formal equality:

$$\textbf{3.3.2} \qquad \det(E(\mathbb{L}_N)) - \det(E(\tilde{\mathbb{L}}_N)) = \omega\det(E_0)$$

We have to prove

$$\textbf{3.2.16} \qquad f(\tilde{\mathbb{L}}) - f(\mathbb{L}) = \frac{\omega q_0 \det(E(\mathbb{L}))}{24}$$

where

$$f(\mathbb{L}) = \frac{q_0}{2}\det(E(\mathbb{L}))\left(s(p_0,q_0) - \frac{p_0}{12q_0} + \frac{\operatorname{sign}(p_0q_0)}{4} \right) - \frac{\det(E(\mathbb{L}_N))}{24q_0}$$

If $p_0 = 0$,

then $q_0 = 1$, and, according to 3.3.2

$$24(f(\tilde{\mathbb{L}}) - f(\mathbb{L})) = q_0\left(\det(E(\mathbb{L}_N)) - \det(E(\tilde{\mathbb{L}}_N))\right) = q_0\omega\det(E_0)$$
$$= \omega q_0 \det(E(\mathbb{L}_{N\cup\{0\}}))$$

So, 3.2.16 is true in this case.

Assume, from now on, that $p_0\neq 0$.

Use the following:

DEDEKIND RECIPROCITY LAW **3.3.3** (See [R-G] Chap. 2):

For any pair $\{p,q\}$ *of nonzero coprime integers:*

$$s(p,q)\operatorname{sign}(q) + s(q,p)\operatorname{sign}(p) = \frac{p^2+q^2+1}{12pq} - \frac{\operatorname{sign}(pq)}{4}$$

❑

to write:

$$f(\mathbb{L}) = \frac{q_0}{2}\det(E(\mathbb{L}))\left(- s(q_0,p_0)\operatorname{sign}(p_0) + \frac{q_0}{12p_0} + \frac{1}{12\,p_0q_0} \right) - \frac{\det(E(\mathbb{L}_N))}{24q_0}$$

So,

3.3.4 $$f(\mathbb{L}) = \frac{q_0}{2}\det(E(\mathbb{L}))\left(-s(q_0,p_0)\mathrm{sign}(p_0) + \frac{q_0}{12p_0}\right) + \frac{\det(E_0)}{24p_0}$$

Similarly, if $q_0+\omega p_0 \neq 0$, we have,

3.3.5

$$f(\tilde{\mathbb{L}}) = \frac{q_0}{2}\det(E(\mathbb{L}))\left(-s(q_0+\omega p_0,p_0)\mathrm{sign}(p_0) + \frac{q_0+\omega p_0}{12p_0}\right) + \frac{\det(\tilde{E}_0)}{24p_0}$$

If $q_0+\omega p_0=0$,

$$f(\tilde{\mathbb{L}}) = -\frac{1}{24p_0}\sum_{\{I \mid I\subset N,\, I\neq\emptyset\}}\det(E(\tilde{\mathbb{L}}_{N\setminus I}))(-1)^{\#I-1}Lk_c(\tilde{L};I\cup\{0\})$$

3.3.6 View $\det(E(\tilde{\mathbb{L}}_{N\setminus I}))$ as the sum running over all permutations σ of $N\setminus I$ of the terms signature(σ)$Lk(\tilde{\mathbb{L}};\sigma)$, and note that the signature of a permutation σ of $N\setminus I$ whose graph is a disjoint union of c oriented circles is

$$\mathrm{signature}(\sigma) = (-1)^{c+n+\#I}$$

This shows that $24p_0f(\tilde{\mathbb{L}})$ is the sum over all the permutations ρ of $N\cup\{0\}$ which do not fix 0 of the terms signature(ρ)$Lk(\tilde{\mathbb{L}};\rho)$, which is $\det(\tilde{E}_0)$.
So, Equality 3.3.5 holds in any case.
Subtracting 3.3.4 from 3.3.5 concludes the proof of 3.2.16.

❑

PROOF OF 3.2.13:
3.2.13 is equivalent to 3.3.7:
3.3.7 If $\det(E(\mathbb{L})) \neq 0$,

$$\mathrm{signature}(E(\tilde{\mathbb{L}})) - \mathrm{signature}(E(\mathbb{L})) = \mathrm{sign}(p_0(q_0+\omega p_0)) - \mathrm{sign}(p_0q_0)$$

With Notation 3.3.1, E_0 and $\tilde{E}_0$ represent the same bilinear symmetric form, and thus have the same signature and the same determinant sign.
Let us recall from 3.2.5 that, if $q_0+p_0\omega = 0$, then $\mathrm{sign}(\tilde{\mathbb{L}}) = p_0\mathrm{sign}(\mathbb{L})$.
Now, 3.3.7 is the consequence of the three following equalities:

3.3.8

$$\mathrm{signature}(E(\mathbb{L})) - \mathrm{signature}(E_0) = \mathrm{sign}\left(\frac{p_0}{q_0}\right)\Big(1 - \mathrm{sign}(\det(E_0)\det(E(\mathbb{L})))\Big)$$

3.3.9 If $q_0+p_0\omega \neq 0$,

$$\mathrm{signature}(E(\tilde{\mathbb{L}})) - \mathrm{signature}(\tilde{E}_0)$$
$$= \mathrm{sign}\left(\frac{p_0}{q_0+p_0\omega}\right)\Big(1 - \mathrm{sign}(\det(\tilde{E}_0)\det(E(\tilde{\mathbb{L}})))\Big)$$

3.3.10 If $q_0+p_0\omega = 0$,

$$\text{signature}(E(\tilde{\mathbb{L}})) - \text{signature}(\tilde{E}_0) = -\,\text{sign}(\det(\tilde{E}_0)\det(E(\tilde{\mathbb{L}})))$$

❑

PROOF OF 3.3.8 (which also proves 3.3.9):
We identify the symmetric matrices with the bilinear symmetric forms they represent.
There exists a codimension 1 subspace, say V, where E_0 (see 3.3.1) and $E(\mathbb{L})$ are equal.
• If E_0 is degenerate, since $E(\mathbb{L})$ is non-degenerate, so is $E(\mathbb{L}_N)$, and

$$\text{signature}(E(\mathbb{L}_N)) = \text{signature}(E_0),$$

furthermore,

$$\text{signature}(E(\mathbb{L})) - \text{signature}(E(\mathbb{L}_N)) = \text{sign}(\frac{\det(E(\mathbb{L}))}{\det(E(\mathbb{L}_N))}) = \text{sign}(\frac{p_0}{q_0})$$

hence,

$$\text{signature}(E(\mathbb{L})) = \text{signature}(E_0) + \text{sign}(\frac{p_0}{q_0})$$

this is 3.3.8 in this case.
• If E_0 is non-degenerate:

Either both E_0 and $E(\mathbb{L})$ are degenerate on V, and they have the same signature and the same determinant sign.

Or both E_0 and $E(\mathbb{L})$ are non-degenerate on V, and the difference between their respective signatures and the signature of their common restriction on V is ± 1.

Hence,

$$|\text{signature}(E(\mathbb{L})) - \text{signature}(E_0)| = \begin{cases} 0 & \text{if } \det(E_0)\det(E(\mathbb{L})) > 0 \\ 2 & \text{if } \det(E_0)\det(E(\mathbb{L})) < 0 \end{cases}$$

Furthermore,

$$\frac{p_0}{q_0}\Big(\text{signature}(E(\mathbb{L})) - \text{signature}(E_0)\Big) \geq 0$$

So, 3.3.8 holds in every case.

❑

PROOF OF 3.3.10:
The proof is immediate, let us just recall that, if $q_0+p_0\omega$ is zero, then:

$$\tilde{\mathbb{L}} = \tilde{\mathbb{L}}_N.$$

❑

This is the end of the proof of 3.2.13. Now, Proposition 3.1.1 is proved except for Equality 3.2.11.

❑ ❑

§3.4 More linking functions: semi-open graphs and functions α

New functions of the linking numbers of L are now needed to express the linking functions of $\tilde{L}$. The particular role played by 0 will be considered.

DEFINITION **3.4.1**: *Semi-open graph*
A *semi-open edge* $E_{i(}$ is a semi-open segment with a vertex v_i ($i \neq 0$) at its end.
A *semi-open graph* with support $I \subset N \cup \{0\}$ is the union of a classical graph with support I and semi-open edges of the form $E_{i(}$ ($i \in I \setminus \{0\}$) attached to the classical graph at v_i.
The *linking of* L *with respect to a semi-open edge* $E_{i(}$ is:

$$Lk(L;E_{i(}) = Lk(K_i,K_0) = \ell_{i0}$$

The *linking* Lk(L;G) *of* L *with respect to a semi-open graph* G is the product over all its (classical and semi-open) edges E of the Lk(L;E).

EXAMPLE AND DEFINITION **3.4.2**: *Lines*
A *line* is a semi-open graph homeomorphic to the real line equipped with an orientation.

Figure 3.1: Two different ways of drawing the same line oriented from left to right with associated linking $\ell_{10}^2 \ell_{20}^2$

Note that a line has always at least one vertex.

NOTATION **3.4.3**: *(Functions* α).
Let I be a subset of N.
Let $E_r(I)$ be the set of semi-open graphs which are disjoint unions of r lines and have support I.
Let f be a real function on $\mathbb{N}$.
Then we set:

$$\alpha(I\,;r \text{—}\,;f(r)) = \sum_{r \in \mathbb{N}} \sum_{G \in E_r(I)} f(r)Lk(L;G)$$

$$(\,\alpha(\emptyset\,;r \text{—}\,;f(r)) = f(0)\,)$$

(Example 3.4.5 will illustrate this definition.)

We introduced semi-open graphs to use the following process:

CUTTING PROCESS **3.4.4**: (Computing the linking number of $\tilde{L}$ with respect to a graph G by *cutting* G.)
Let G be a graph with support in $N \cup \{0\}$. Let $\mathfrak{C}(G)$ denote the set of subsets of the set of the edges of G which do not contain the vertex v_0.
For $C \in \mathfrak{C}(G)$, let G_C be the semi-open graph obtained from G by cutting each edge of C, that is by removing a point from the interior of each edge of C. This operation of removing a point is called a *cut,* and C is called a *set of cuts.*
Then:

$$Lk(\tilde{L};G) = \sum_{C \in \mathfrak{C}(G)} Lk(L;G_C)\omega^{\#C}$$

(This is nothing more than replacing $\tilde{\ell}_{ij}$ (if $i,j \neq 0$) by $(\ell_{ij} + \omega \ell_{i0}\ell_{j0})$ and expanding.)
A cut will be symbolized by a big grey point on the cut edge as shown in Figures 3.2 and 3.3.

EXAMPLE **3.4.5**: The cutting process can be used to prove the following equalities (which will be useful in §3.5):
Let I *be a subset of* N, $n \notin I$, *then:*

3.4.6

$$Lk_c(\tilde{L}_{I \cup \{n\}}) = Lk_c(L_{I \cup \{n\}}) + \sum_{\{J \,/\, J \subset I\}} \left(\alpha(I \setminus J\,;\, r — ;\, r!\, \omega^{r+1})\, Lk_c(L_{J \cup \{0,n\}}) \right)$$

3.4.7 $$Lk_c(\tilde{L}_{I \cup \{0,n\}}) = \sum_{\{J \,/\, J \subset I\}} \left(\alpha(I \setminus J\,;\, r — ;\, (r+1)!\omega^{r})\, Lk_c(L_{J \cup \{0,n\}}) \right)$$

PROOF OF 3.4.6:
$Lk_c(\tilde{L}_{I \cup \{n\}})$ is the sum running over all pairs $(\mathfrak{T},C)$ where

- $\mathfrak{T}$ is an oriented circle with support $I \cup \{n\}$,
- $C \in \mathfrak{C}(\mathfrak{T})$ is a set of cuts of $\mathfrak{T}$,

of the terms $Lk(L;\mathfrak{T}_C)\omega^{\#C}$.
The restriction of this sum to the pairs $(\mathfrak{T}, \emptyset)$ is $Lk_c(L_{I \cup \{n\}})$.
The sum involving the functions α in the right-hand side of 3.4.6 is the sum running over all pairs (G,O) where

- G is a disjoint union of r lines and an oriented circle with 0 and n in its support, the support of G is $I \cup \{0,n\}$,
- O is an order on the r lines of G,

of the terms $Lk(L;G)\omega^{r+1}$. (The order O allows us to forget r!.)
We establish a one-to-one correspondence between the $(\mathfrak{T},C)$, $(C \neq \emptyset)$, and the (G,O) as shown in Figure 3.2:

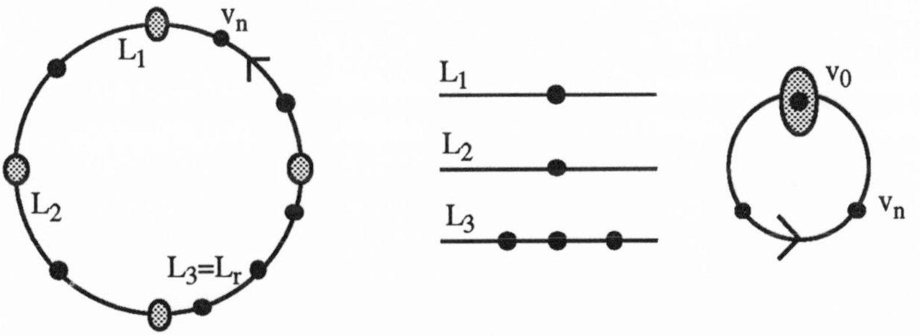

Figure 3.2: Proving 3.4.6 by cutting oriented circles with support $I\cup\{n\}$

Starting with a pair $(\mathcal{C}, C)$, $(C\neq\emptyset)$, we first cut $\mathcal{C}$, according to C (that is removing the grey points in Figure 3.2), to get $(\#C)$ lines.
Then, we glue the two "ends" of the line containing v_n on a new vertex v_0 in order to obtain an oriented circle with 0 and n in its support. (The orientation of the circle is induced by the orientation of $\mathcal{C}$.) We obtain thus a semi-open graph $G(\mathcal{C},C)$ and an order $O(\mathcal{C},C)$ induced on the $(r(G)=\#C-1)$ lines of $G(\mathcal{C},C)$ by the orientation of $\mathcal{C}$. The pair $(G(\mathcal{C},C),O(\mathcal{C},C))$ satisfies

$$Lk(L;G)\omega^{r(G)+1} = Lk(L;\mathcal{C}_C)\omega^{\#C}$$

Since the correspondence is bijective, Equality 3.4.6 is satisfied.

❑

PROOF OF 3.4.7:

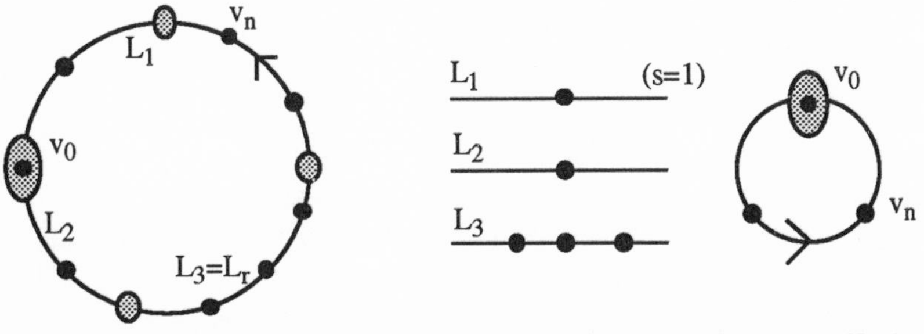

Figure 3.3: Proving 3.4.7 by cutting oriented circles with support $I\cup\{0,n\}$

The principle is the same.
Here we will view the sum involving the functions α as the sum running over all triples (G,O,s) where

- G is a disjoint union of r lines and of an oriented circle with 0 and n in its support, the support of G is $I\cup\{0,n\}$,
- O is an order on the r lines of G,

• s is a number between 0 and r,

of the terms $Lk(L;G)\omega^r$.

Thus, there are (r+1)! pairs (O,s) associated with G; and s is used, when defining the one-to-one correspondence, to locate v_0 on the cut circle $\mathfrak{C}_C$ with support $I\cup\{0,n\}$ reconstructed from (G,O,s): s counts the cuts from v_n to v_0 on $\mathfrak{C}$.

□

§3.5 Variation of the ζ-coefficients under an ω-twist

This subsection shows the following proposition relating the ζ-coefficients of the sublinks of $\tilde{L}$ to the ζ-coefficients of the sublinks of L.

PROPOSITION:

For any subset I *of* N, *Relation* 3.5.1 *holds:*

3.5.1

$$\zeta_{N\cup\{0\}}(\tilde{L}_{I\cup\{0\}}) = \sum_{\{J\,/\,J\subset I\}}\left((-1)^{\#I+\#J}\,\alpha(I\setminus J\,;\ r\,—\,;\ (r+1)!\omega^r)\ \zeta_{N\cup\{0\}}(L_{J\cup\{0\}})\right)$$

For any non-empty subset I *of* N, *Relation* 3.5.2 *holds:*

3.5.2

$$\zeta_{N\cup\{0\}}(\tilde{L}_I) = \zeta_{N\cup\{0\}}(L_I) - \omega \sum_{\{J\,/\,J\subset I\}}\left((-1)^{\#I+\#J}\,\alpha(I\setminus J\,;\ r\,—\,;\ r!\ \omega^r)\ \zeta_{N\cup\{0\}}(L_{J\cup\{0\}})\right)$$

□

Recall that N is the set of the n indices of the components of L_N. To simplify notation, we often view it as a subset of $\mathbb{N}$. This induces an order on the elements of N. This order affects none of the given functions. (The only thing which matters is the correspondence between the indices and the components.) In this chapter (unless otherwise mentioned), N is supposed to be fixed, but it is arbitrary. In particular, if a relation holds for N (without restrictive assumptions), it holds for any subset of N.

To prove the proposition, we will prove Assertions 3.5.3 to 3.5.7:

3.5.3 *For any subset* I *of* N, *Relation* 3.5.1 *holds for* I *if and only if the following relation holds for* I:

$$\zeta_{I\cup\{0\}}(\tilde{L}_{I\cup\{0\}}) = \sum_{\{J\,/\,J\subset I\}}\left((-1)^{\#I+\#J}\,\alpha(I\setminus J\,;\ r\,—\,;\ (r+1)!\omega^r)\ \zeta_{I\cup\{0\}}(L_{J\cup\{0\}})\right)$$

3.5.4 *For any non-empty subset* I *of* N, *Relation* 3.5.2 *holds for* I *if and only if the following relation holds for* I:

$$\zeta_{I\cup\{0\}}(\tilde{L}_I) = \zeta_{I\cup\{0\}}(L_I) - \omega \sum_{\{J\,/\,J\subset I\}}\left((-1)^{\#I+\#J}\,\alpha(I\backslash J\,;\,r\mapsto\,;\,r!\,\omega^r)\,\zeta_{I\cup\{0\}}(L_{J\cup\{0\}})\right)$$

3.5.5 *Relation* 3.5.1 *holds if* I = N.

3.5.6 *If Relation* 3.5.2 *holds for any non-empty subset* I *strictly included in* N, *and if* ℓ_{10} *is nonzero, then Relation* 3.5.2 *holds for* I = N.

LEMMA **3.5.7**: *Relation* 3.5.2 *is true if* I=N *and if* ℓ_{i0} *is zero for any element* i *of* I.

❑

PROOF OF THE PROPOSITION (using Assertions 3.5.3 to 3.5.7 which will be proved later)

According to the remarks following the statement of the proposition, 3.5.1 follows from 3.5.3 and 3.5.5, while 3.5.2 follows from 3.5.4, 3.5.6 and 3.5.7 by induction on the nonzero cardinality of N.

❑

PROOF OF 3.5.3 AND 3.5.4

To prove 3.5.3 and 3.5.4, respectively, it suffices to prove:

For any subset I of N, and for any element n of N\I:

3.4.7 $$Lk_c(\tilde{L}_{I\cup\{0,n\}}) = \sum_{\{J\,/\,J\subset I\}}\left(\alpha(I\backslash J\,;\,r\mapsto\,;\,(r+1)!\,\omega^r)\,Lk_c(L_{J\cup\{0,n\}})\right)$$

3.4.6

$$Lk_c(\tilde{L}_{I\cup\{n\}}) = Lk_c(L_{I\cup\{n\}}) + \sum_{\{J\,/\,J\subset I\}}\left(\alpha(I\backslash J\,;\,r\mapsto\,;\,r!\,\omega^{r+1})\,Lk_c(L_{J\cup\{0,n\}})\right)$$

Since these equalities have already been proved in Example 3.4.5, we are done.

❑

PRELIMINARIES FOR THE PROOFS OF 3.5.5 AND 3.5.6:

The ω-twist of Definition 1.6.4 is a homeomorphism from M\L onto M\$\tilde{L}$.

Assume (without loss of generality) that L and $\tilde{L}$ are oriented in a consistent way.

Then, the oriented meridians of $\tilde{L}$, $\tilde{m}_0, \tilde{m}_1, \ldots, \tilde{m}_n$, are related to the oriented meridians of L, $m_0, m_1, \ldots, m_n$ by the following positive change of basis:

3.5.8 $$\begin{cases} \tilde{m}_0 = m_0 - \sum_{i=1}^{n} \omega \ell_{i0} m_i & \\ \tilde{m}_i = m_i & \text{if } i \geq 1 \end{cases}$$

So, if we write the Alexander series of L as

$$\textbf{3.5.9} \qquad \mathcal{D}(L)(u_0,u_1,\ldots,u_n) = \sum_{K=(k_0,k_1,\ldots,k_n)\in\mathbb{Q}^{n+1}} a_K \exp\left(\sum_{i=0}^{n} k_i u_i\right)$$

according to 2.3.2, the Alexander series of $\tilde{L}$ can be written as follows:

$$\textbf{3.5.10} \qquad \mathcal{D}(\tilde{L})(\tilde{u}_0,u_1,\ldots,u_n) = \sum_K a_K \exp\left(k_0\tilde{u}_0 + \sum_{i=1}^{n}(k_i + k_0\omega\ell_{i0})u_i\right)$$

PROOF OF 3.5.5 (or 3.5.1 for I=N):
According to 3.5.10,
3.5.11

$$\mathcal{D}_{\mathbb{1}}(\tilde{L}) = \frac{\partial^{n+1}}{\partial\tilde{u}_0\partial u_1\ldots\partial u_n}\mathcal{D}(\tilde{L})(0) = \sum_K a_K k_0 \prod_{i=1}^{n}(k_i+\omega\ell_{i0}k_0)$$

$$= \sum_{\{R\,/\,R \text{ multiindex},\ S(R)=N\cup\{0\},\ |R|=n+1,\ r_i\le 1\ \forall i\ge 1\}} \left(\sum_K a_K K^R\right)\omega^{r_0-1}\prod_{i=1}^{n}\ell_{i0}^{1-r_i}$$

So, according to 3.5.9,

$$\textbf{3.5.12} \qquad \mathcal{D}_{\mathbb{1}}(\tilde{L}) = \sum_{\{R\,/\,R \text{ multiindex},\ S(R)=N\cup\{0\},\ |R|=n+1,\ r_i\le 1\ \forall i\ge 1\}} \mathcal{D}_R(L)\, r_0!\, \omega^{r_0-1}\prod_{i=1}^{n}\ell_{i0}^{1-r_i}$$

Replacing the $\mathcal{D}_R(L)$ by their expressions given in 2.5.2 leads to:

$$\textbf{3.5.13} \qquad \zeta_{N\cup\{0\}}(\tilde{L}_{N\cup\{0\}}) = (-1)^n \sum_{\{J\,/\,J\subset N\}} \gamma_J\, \zeta_{N\cup\{0\}}(L_{J\cup\{0\}})$$

with

$$\gamma_J = (-1)^{\#J} \sum_{\substack{\{R\,/\,R \text{ multiindex},\ S(R)=N\cup\{0\},\\ |R|=n+1,\ r_i\le 1\ \forall i\ge 1,\ r_j=1\ \forall j\in J\}}} Lk(L;N\setminus J, \to J\cup\{0\};R)\, r_0!\, \omega^{r_0-1}\prod_{i=1}^{n}\ell_{i0}^{1-r_i}$$

A functional graph appearing in $Lk(L;N\setminus J, \to J\cup\{0\};R)$ is a wedge W of (r_0-1) segments going to v_0 as shown in the left-hand side of Figure 3.4 . We can first add semi-open edges to the origins of these segments to take $\prod_{i=1}^{n}\ell_{i0}^{1-r_i}$ into account and next remove a neighborhood of v_0 to transform W into a disjoint union of $r=(r_0-1)$ lines as shown in the right-hand side of Figure 3.4.

Figure 3.4

So,

$$\textbf{3.5.14} \qquad \gamma_J = (-1)^{\#J}\,\alpha(N\backslash J\,;\ r\,—\,;\ (r+1)!\omega^r)$$

This proves Equality 3.5.1 when I = N, that is 3.5.5.

❑

PROOF OF 3.5.6

We have to prove 3.5.2 under the assumptions:

- $I = N$
- $\ell_{10} \neq 0$
- Relation 3.5.2 holds for any non-empty proper subset I of N. (Induction hypothesis)

Let S be the multiindex $S = (s_0, s_1, \ldots, s_n)$ with $s_0 = 0$, $s_1 = 2$, and $s_i = 1$ for $i \geq 2$.

Equality 3.5.2 will be checked by computing $\mathcal{D}_S(\tilde{L})$ in two different ways:

First computation of $\mathcal{D}_S(\tilde{L})$:

According to 3.5.10 and 3.5.9,

$$2\mathcal{D}_S(\tilde{L}) = \sum_K a_K\,(k_1+\omega\ell_{10}k_0)^2\prod_{i=2}^{n}(k_i+\omega\ell_{i0}k_0)$$

$$= \sum_{\{R\,/\,R \text{ multiindex},\ S(R) = N\cup\{0\},\ |R| = n+1,\ r_i\leq s_i\ \forall i\geq 1\}} \mathcal{D}_R(L)\ r_0!\,\frac{2}{(s_1-r_1)!}\,\omega^{r_0}\prod_{i=1}^{n}\ell_{i0}^{s_i-r_i}$$

So, according to 2.5.2,

$$\textbf{3.5.15} \qquad \mathcal{D}_S(\tilde{L}) = (-1)^{n+1} \sum_{\{I\,/\,I\subset N\cup\{0\},\ \{0,1\}\cap I\neq\emptyset\}} \beta_I\,\zeta_{N\cup\{0\}}(L_I)$$

where:

3.5.16

$$\beta_I = (-1)^{\#I} \sum_{\substack{\{R \,/\, R \text{ multiindex, } S(R) = N\cup\{0\},\\ |R| = n+1,\ r_i \leq s_i\ \forall i\geq 1,\ r_i\geq 1\ \forall i\in I\}}} Lk(L;N\cup\{0\}\backslash I, \to I;R)\, \frac{r_0!}{(s_1-r_1)!}\, \omega^{r_0} \prod_{i=1}^{n} \ell_{i0}^{s_i-r_i}$$

Second computation of $\mathcal{D}_S(\tilde{L})$:

Proposition 2.5.2 gives:

$$\mathcal{D}_S(\tilde{L}) = (-1)^{n+1} \sum_{\{I \,/\, I\subset N,\ 1\in I\}} Lk(\tilde{L};N\cup\{0\}\backslash I,\to I;S)\,(-1)^{\#I}\zeta_{N\cup\{0\}}(\tilde{L}_I)$$

Using the induction hypothesis leads to:

3.5.17

$$\begin{aligned}\mathcal{D}_S(\tilde{L}) &= (-1)^{n+1} \sum_{\{I \,/\, I\subset N,\ 1\in I\}} Lk(\tilde{L};N\cup\{0\}\backslash I,\to I;S)\,(-1)^{\#I}\zeta_{N\cup\{0\}}(L_I)\\ &\quad - (-1)^{n+1}\omega \sum_{\{J \,/\, J\subset N\}} \gamma_J\, \zeta_{N\cup\{0\}}(L_{J\cup\{0\}})\\ &\quad - \ell_{10}\Big(\zeta_{N\cup\{0\}}(\tilde{L}_N) - \zeta_{N\cup\{0\}}(L_N)\Big)\\ &\quad - \omega\ell_{10} \sum_{\{J \,/\, J\subset N\}} (-1)^{n+\#J}\, \alpha(N\backslash J\,;\, r \longrightarrow\,;\, r!\,\omega^r)\, \zeta_{N\cup\{0\}}(L_{J\cup\{0\}})\end{aligned}$$

where

3.5.18

$$\gamma_J = \sum_{\{I \,/\, J\cup\{1\}\subset I\subset N\}} \Big((-1)^{\#J}\, \alpha(I\backslash J\,;\, r \longrightarrow\,;\, r!\,\omega^r)\, Lk(\tilde{L};N\cup\{0\}\backslash I,\to I;S)\Big)$$

To prove Equality 3.5.2 (for I=N), it suffices to check that the sum of the two last lines in 3.5.17 is zero. So, according to 3.5.15 and 3.5.17, it suffices to verify the following assertions:

3.5.19 For any subset I of N which contains 1,

$$\beta_I = (-1)^{\#I}\, Lk(\tilde{L};N\cup\{0\}\backslash I,\to I;S)$$

3.5.20 For any subset J of N,

$$\beta_{J\cup\{0\}} = -\,\omega\gamma_J$$

3.5.21 To prove these equalities, we will again write both of their sides as sums running over graphs and cut graphs equipped with structures, and we will establish one-to-one correspondences between identical elementary summands. We will always *complete* the functional graphs involved in the β-coefficients (see 3.5.16) by adding to them (s_i-r_i) semi-open edges attached to the vertices v_i, for $i \in N$, so that the linking of L with respect to the semi-open *completed* graph G of a function f of $\mathfrak{F}(N\cup\{0\}\backslash I,\rightarrow I;R)$ satisfies

$$Lk(L;G) = Lk(L;f)\prod_{i=1}^{n} \ell_{i0}^{s_i-r_i}$$

PROOF OF 3.5.19:

For any subset I of N which contains 1, we have to prove:

$$Lk(\tilde{L};N\cup\{0\}\backslash I,\rightarrow I;S)$$

$$= \sum_{\substack{\{R \,/\, R \text{ multiindex, } S(R) = N\cup\{0\},\\ |R| = n+1,\ r_i \le s_i\ \forall i\ge 1,\ r_i \ge 1\ \forall i\in I,\ r_1=2\}}} Lk(L;N\cup\{0\}\backslash I, \rightarrow I;R)\,\frac{r_0!}{(s_1-r_1)!}\,\omega^{r_0}\prod_{i=1}^{n} \ell_{i0}^{s_i-r_i}$$

A functional graph involved in $Lk(\tilde{L};N\cup\{0\}\backslash I,\rightarrow I;S)$ equipped with r_0 cuts is a cut segment from v_0 to v_1 with support $(N\cup\{0\}\backslash I)\cup\{1\}$, as shown in the left-hand side of Figure 3.5. It corresponds to one completed functional graph (see 3.5.21) involved in $Lk(L;N\cup\{0\}\backslash I, \rightarrow I;R)$ (where R is determined by the cuts of the segment) equipped with one of the $r_0!$ possible orders of the r_0 branches which go to v_0 (see the right-hand side of Figure 3.5).

This proves 3.5.19.

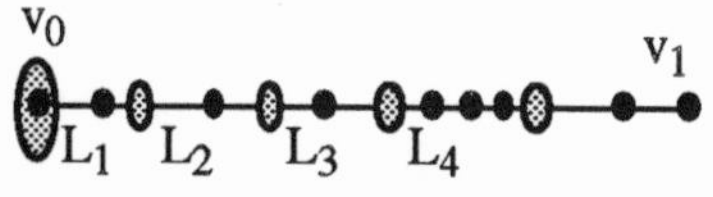

Cut functional graph involved in $Lk(\tilde{L};N\cup\{0\}\backslash I,\rightarrow I;S)$

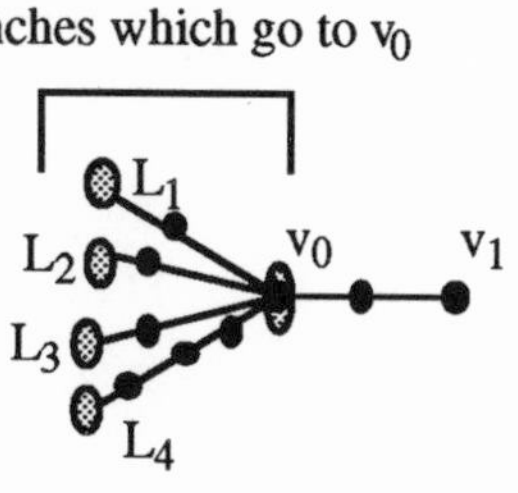

Corresponding completed functional graph involved in $Lk(L;N\cup\{0\}\backslash I, \rightarrow I;R)$

Figure 3.5

❑

PROOF OF 3.5.20:

The equality to be proved for any subset J of N is:

3.5.22
$$\sum_{\{I \,/\, J\cup\{1\} \subset I \subset N\}} \left(\alpha(I\setminus J \,;\, r — \,;\, r!\ \omega^{r+1})\ Lk(\tilde{L};N\cup\{0\}\setminus I,\to I;S) \right)$$

$$= \sum_{\substack{\{R \,/\, R \text{ multiindex},\ S(R) = N\cup\{0\},\\ |R| = n+1,\ r_i\leq s_i\ \forall i\geq 1,\ r_i\geq 1\ \forall i\in J\cup\{0\}\}}} Lk(L;N\setminus J, \to J\cup\{0\};R)\, \frac{r_0!}{(s_1-r_1)!}\, \omega^{r_0} \prod_{i=1}^{n} \ell_{i0}^{s_i-r_i}$$

Case 1: $1\in J$

A typical graph involved in $\left(\alpha(I\setminus J \,;\, r — \,;\, r!\ \omega^{r+1})\ Lk(\tilde{L};N\cup\{0\}\setminus I,\to I;S)\right)$ is the union of a disjoint ordered union of r lines with support $I\setminus J$ and of a cut segment from v_0 to v_1. It corresponds to a completed functional graph involved in a well-determined $Lk(L;N\setminus J,\to J\cup\{0\};R)$ equipped with an order on the $(r_0\text{-}1)$ segments which go to v_0 and with a number r between 0 and $(r_0\text{-}1)$.
The number r corresponds to the number of lines in the partition of $I\setminus J$. See Figure 3.6.

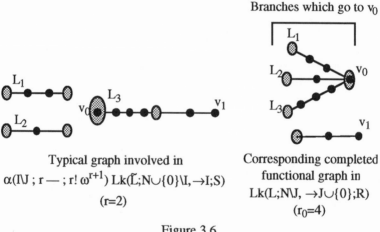

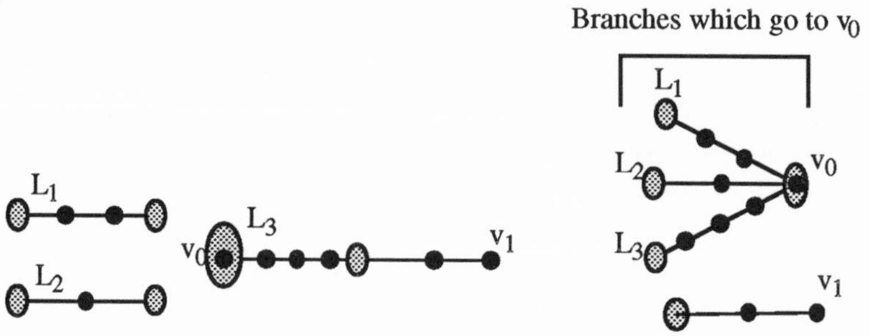

Typical graph involved in $\alpha(I\setminus J \,;\, r — \,;\, r!\ \omega^{r+1})\ Lk(\tilde{L};N\cup\{0\}\setminus I,\to I;S)$ $(r=2)$

Corresponding completed functional graph in $Lk(L;N\setminus J, \to J\cup\{0\};R)$ $(r_0=4)$

Figure 3.6

This proves 3.5.20 when 1 belongs to J.

❑

Case 2: $1\notin J$

In a completed functional graph involved in $Lk(L;N\setminus J, \to J\cup\{0\};R)$ (in 3.5.22), exactly two semi-open segments go to v_1. (A semi-open edge $E_{1(}$ is a segment.) One of these two segments will be labelled as the higher segment going to v_1, the other one as the lower one (see Figure 3.7). So, from one typical graph involved in Lk(L;N\J, $\rightarrow$J$\cup$\{0\};R), we get one labelled graph if r_1=0 and two if $r_1\neq 0$. The right-hand side of 3.5.22 is now the sum over such labelled completed graphs G of the terms $\frac{r_0!}{2}\,\omega^{r_0}$ Lk(L;G).

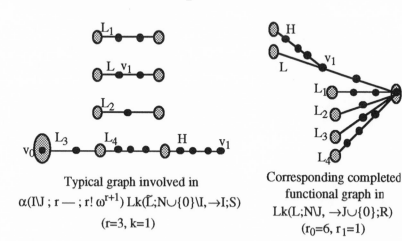

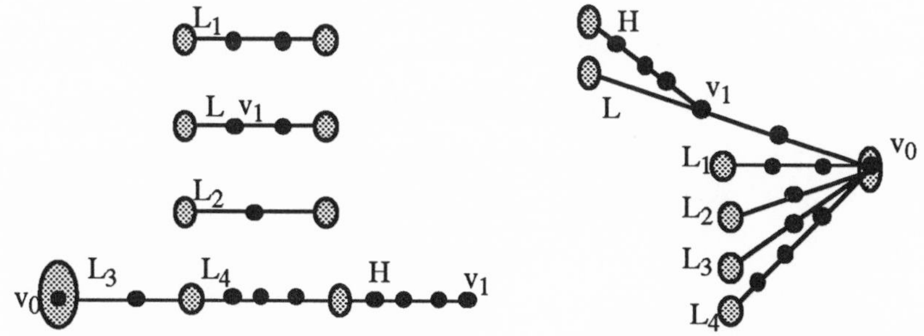

Typical graph involved in
α(I\J ; r — ; r! ω^{r+1}) Lk($\tilde{L}$;N$\cup$\{0\}\I,$\rightarrow$I;S)
(r=3, k=1)

Corresponding completed functional graph in
Lk(L;N\J, $\rightarrow$J$\cup$\{0\};R)
(r_0=6, r_1=1)

Figure 3.7

An elementary cut graph in $\Big($ α(I\J ; r — ; r! ω^{r+1}) Lk($\tilde{L}$;N$\cup$\{0\}\I,$\rightarrow$I;S)$\Big)$ equipped with an order for the segments in the α-term corresponds now to a labelled functional graph of Lk(L;N\J, $\rightarrow$J$\cup$\{0\};R) equipped with one of the $\frac{r_0!}{2}$ structures consisting of:

- an order on the (r_0-2) segments which go to v_0 and do not contain v_1,
- a number r between 1 and (r_0-1), and
- a number k between 0 and (r-1) counting the segments occurring before the segment containing v_1 among the r segments in the α-term.

This proves 3.5.20 when 1 does not belong to J.

❑

3.5.6 is checked.

❑ ❑

PROOF OF LEMMA 3.5.7

This lemma can be written as follows.

If, for any element i *of* N, ℓ_{i0} *is zero, then:*

$$\textbf{3.5.23}\qquad \zeta(\tilde{L}_N) = \zeta(L_N) - \omega\,\zeta(L_{N\cup\{0\}})$$

PROOF OF 3.5.23

ϕ denotes the map

$$\phi\colon \mathbb{Z}[\exp(\pm\frac{u_i}{2O_M(K_i)})]_{i=0,\ldots,n} \to \mathbb{Z}[\exp(\pm\frac{u_i}{2O_M(K_i)})]_{i=1,\ldots,n}$$

obtained by setting $u_0 = 0$.

We first note that, since the ℓ_{i0} are zero, the three following natural compositions are the same:

$$\mathbb{Z}[\frac{1}{2}Q_{fa}(M\setminus L)] \to \mathbb{Z}[\frac{1}{2}Q_{fa}(M\setminus L_N)] \overset{\psi_{L_N}}{\to} \mathbb{Z}[\exp(\pm\frac{u_i}{2O_M(K_i)})]_{i=1,\ldots,n}$$

$$\mathbb{Z}[\frac{1}{2}Q_{fa}(M\setminus L)] \to \mathbb{Z}[\frac{1}{2}Q_{fa}(M\setminus \tilde{L}_N)] \overset{\psi_{\tilde{L}_N}}{\to} \mathbb{Z}[\exp(\pm\frac{u_i}{2O_M(K_i)})]_{i=1,\ldots,n}$$

$$\mathbb{Z}[\frac{1}{2}Q_{fa}(M\setminus L)] \overset{\psi_L}{\to} \mathbb{Z}[\exp(\pm\frac{u_i}{2O_M(K_i)})]_{i=0,\ldots,n}$$

$$\overset{\phi}{\to} \mathbb{Z}[\exp(\pm\frac{u_i}{2O_M(K_i)})]_{i=1,\ldots,n}$$

(See 2.2.1 and 3.5.8.)

Next, we compare the Alexander series $\mathcal{D}(L)$, $\mathcal{D}(L_N)$ and $\mathcal{D}(\tilde{L}_N)$ computed from Definition 2.2.2 and from convenient CW-complexes, strong deformation retracts of $M\setminus L$, $M\setminus L_N$ and $M\setminus \tilde{L}_N$:

We use a CW-complex X, strong deformation retract of $M\setminus L$, with:

- the basepoint of $M\setminus L$ as a unique zero-cell,
- (k+1) one-cells, $c_0, c_1, \ldots, c_k$, such that $c_0, c_1, \ldots, c_{n-1}$ are the oriented meridians of $K_0, K_1, \ldots, K_{n-1}$, the last cell c_k is the oriented meridian of K_n, and c_n is the longitude of K_0 which bounds a disk in $M\setminus L_{\{0\}}$,
- k two-cells, $e_0, e_1, \ldots, e_{k-1}$, such that (the closures of) the cells $e_0, e_1, \ldots, e_{n-1}$ are the boundaries of the tubular neighborhoods of $K_0, K_1, \ldots, K_{n-1}$.

(e_0 is attached to the 1-skeleton X^1 of X along $c_0c_nc_0^{-1}c_n^{-1}$.)

For $M\setminus L_N$ and $M\setminus \tilde{L}_N$, we use the CW-complexes obtained from the previous one by replacing e_0 by e for $M\setminus L_N$, and by $\tilde{e}$ for $M\setminus \tilde{L}_N$, where e and $\tilde{e}$ are attached to X^1 along c_0 and $c_n^{-\omega}c_0$, respectively.

We use the same notation for a cell and for a preferred lifting of this cell in the maximal free abelian covering of its ambient space Y; and we call ∂ the map which sends (liftings of) two-cells to their boundaries in the cellular $\mathbb{Z}[Q_{fa}(Y)]$-complex associated with Y. (See Remark 2.2.3.)

Then (for appropriate lifting choices), since c_n is null-homologous in $M\setminus L$:

$$\partial(e_0) = (\psi_L{}^{-1}(\exp(u_0) - 1))c_n$$
$$\partial(e) = c_0$$

and

$$\partial(\tilde{e}) = c_0 - \omega c_n$$

(See Example A.2.27.)

We denote by

$$\partial_{ij} \in \mathbb{Z}[Q_{fa}(M\setminus L)]$$

the coordinate of $\partial(e_j)$ with respect to c_i in the $\mathbb{Z}[Q_{fa}(M\setminus L)]$-complex associated with X.
To get the right normalizations, we assume $\varepsilon(\det([\partial_{ij}]_{n\leq i,j\leq k-1})) > 0$.

Then, according to Definition 2.2.2, there exist positive units

$$U_0 \in \mathbb{Z}[\exp(\pm\frac{u_i}{2O_M(K_i)})]_{i=0,\ldots,n}$$

and

$$U, \tilde{U} \in \mathbb{Z}[\exp(\pm\frac{u_i}{2O_M(K_i)})]_{i=1,\ldots,n}$$

such that:
3.5.24

$$(\exp(u_n) - 1)\mathcal{D}(L) = U_0\,(\exp(u_0) - 1)\,(-1)^n\,\psi_L(\det(D_n))$$
$$(\exp(u_n) - 1)\mathcal{D}(L_N) = U\,\phi\circ\psi_L(\det(D_0)) \qquad \text{(if } n > 1)$$
$$(\exp(u_n) - 1)\mathcal{D}(\tilde{L}_N) = \tilde{U}\,\phi\circ\psi_L(\det(D_0) - \omega\,(-1)^n\det(D_n)) \qquad \text{(if } n > 1)$$

where

$$D_0 = [\partial_{ij}]_{0<i,j\leq k-1} \text{ and } D_n = [\partial_{ij}]_{0\leq i,j\leq k-1,\, i\neq n,\, j\neq 0}$$

Hence, if $n > 1$, there exist positive units

$$U_1, U_2 \in \mathbb{Z}[\exp(\pm\frac{u_i}{2O_M(K_i)})]_{i=1,\ldots,n}$$

such that:

$$\textbf{3.5.25} \qquad \mathcal{D}(\tilde{L}_N) = U_1\mathcal{D}(L_N) - \omega\,U_2\,\phi\Big(\frac{\mathcal{D}(L)}{\exp(\frac{u_0}{2}) - \exp(-\frac{u_0}{2})}\Big)$$

So, according to the properties of the first terms of the Alexander series (see 2.5.2), the derivative of 3.5.25 at 0 with respect to the multiindex

$$\mathbb{1}_N = \mathbb{1} = (1,\ldots,1)$$

is

$$\zeta(\tilde{L}_N) = \zeta(L_N) - \omega\,\zeta(L) + \frac{1}{2}\sum_{i=1}^{n}\sum_{j\in N\setminus\{i\}}\frac{\partial^2}{\partial u_i\partial u_j}(U_1)(0)\mathcal{D}_{\mathbb{1}_{ij}}(L_N)$$

Now, taking the derivative of 3.5.25 at 0 with respect to $\mathbb{1}_i$ leads to

$$\sum_{j \in N \setminus \{i\}} \frac{\partial}{\partial u_j}(U_1)(0) \mathcal{D}_{\mathbb{1}_{ij}}(L_N) = 0,$$

where, since U_1 is a positive unit,

$$\frac{\partial^2}{\partial u_i \partial u_j}(U_1)(0) = \frac{\partial}{\partial u_i}(U_1)(0) \frac{\partial}{\partial u_j}(U_1)(0)$$

This proves 3.5.23 if $n > 1$.

If $n = 1$, $(\exp(u_n) - 1)$ must be replaced by $\dfrac{\exp(u_1) - 1}{\exp(\frac{u_1}{O_M(K_1)}) - 1}$ in the last two equalities of 3.5.24, and 3.5.25 must be replaced by:

3.5.26

$$\mathcal{D}(\tilde{L}_N) = U_1 \mathcal{D}(L_N) - \omega\, U_2\, \phi\left(\frac{\exp(\frac{u_1}{2O_M(K_1)}) - \exp(- \frac{u_1}{2O_M(K_1)})}{\exp(\frac{u_0}{2}) - \exp(- \frac{u_0}{2})} \mathcal{D}(L) \right)$$

Here, taking the derivative of 3.5.26 at 0 with respect to u_1 shows that U_1 (and hence U_2) equals 1 in 3.5.26. So, 3.5.23 is true for any n, and Lemma 3.5.7 is proved.

❑

So are 3.5.1 and 3.5.2.

❑ ❑

§3.6 Proof of 3.2.11 (variation of the piece containing the ζ-coefficients under the ω-twist)

This last subsection completes the proof of Proposition 3.1.1 by proving 3.2.11:

3.2.11
$$a(\tilde{\mathbb{L}}) - a(\mathbb{L}) = \frac{|H_1(M)|\, \omega q_0 \det(E(\mathbb{L}_{N \cup \{0\}}))}{24} \left(\sum_{i=1}^{n} \ell_{i0}^2 - 1 \right)$$

where

$$a(\tilde{\mathbb{L}}) = \sum_{\{ I \,/\, I \neq \emptyset ,\, I \subset N \cup \{0\} \}} (q_0 + p_0 \omega) \det(E(\tilde{\mathbb{L}}_{N \cup \{0\} \setminus I})) \zeta_{N \cup \{0\}}(\tilde{L}_I)$$

and

$$a(\mathbb{L}) = \sum_{\{ I \,/\, I \neq \emptyset ,\, I \subset N \cup \{0\} \}} q_0 \det(E(\mathbb{L}_{N \cup \{0\} \setminus I})) \zeta_{N \cup \{0\}}(L_I)$$

According to 3.2.3 and to Equalities 3.5.1 and 3.5.2 of the previous subsection, we have:

3.6.1

$$a(\tilde{\mathbb{L}}) - a(\mathbb{L}) = \sum_{\{J / J \subset N\}} \Big(u(J) - q_0 \det(E(\mathbb{L}_{N\setminus J})) - v(J) \Big) \zeta_{N\cup\{0\}}(L_{J\cup\{0\}})$$

where

$u(J) =$

$$\sum_{\{I / J \subset I \subset N\}} \Big((-1)^{\#I+\#J} \alpha(I\setminus J \;;\; r — \;;\; (r+1)!\omega^r) (q_0+p_0\omega)\det(E(\tilde{\mathbb{L}}_{N\setminus I})) \Big)$$

$v(J) =$

$$\omega \sum_{\{I / J \subset I \subset N,\, I \neq \emptyset\}} \Big((-1)^{\#I+\#J} \alpha(I\setminus J \;;\; r — \;;\; r!\omega^r) q_0 \det(E(\mathbb{L}_{N\cup\{0\}\setminus I})) \Big)$$

Since

$$\zeta_{N\cup\{0\}}(L_{\{0\}}) = \frac{|H_1(M)|}{24} \left(\sum_{i=1}^{n} \ell_{i0}^2 - 1 \right)$$

to prove 3.2.11, it suffices to prove:

3.6.2

$$u(J) - q_0 \det(E(\mathbb{L}_{N\setminus J})) - v(J) = \begin{cases} = 0 & \text{if } J \neq \emptyset \\ = \omega q_0 \det(E(\mathbb{L}_{N\cup\{0\}})) & \text{if } J = \emptyset \end{cases}$$

To prove 3.6.2 we need to extend naturally the functions α introduced in 3.4.3.

NOTATION **3.6.3**: More functions α

If f is a function on $\mathbb{N}\times\mathbb{N}$ and if $\mathcal{P}(r)$ is a property of an integer r,

$$\alpha(N\setminus J \;;\; r — [\mathcal{P}(r)];\; s\,\mathrm{O} \;;\; f(r,s))$$

represents the sum running over the triples (r,s,G) where:

- $(r,s) \in \mathbb{N}\times\mathbb{N}$, and $\mathcal{P}(r)$ is satisfied,
- G is a semi-open graph with support $N\setminus J$, G is the union of r lines and of s oriented circles,

of the terms f(r,s)Lk(L;G).

EXAMPLES:

3.6.4 $\quad \alpha(N \;;\; r — [r=1];\; s\,\mathrm{O} \;;\; (-1)^{n+s}) = \det(E(\mathbb{L}_{N\cup\{0\}})) - \dfrac{p_0}{q_0} \det(E(\mathbb{L}_N))$

Indeed, we can attach the two "ends" of the line involved in the argument of α to a new vertex v_0 to get an oriented circle. This transforms a graph involved in α into the graph of a permutation σ of $N\cup\{0\}$ which does not fix 0; and α is the sum, over such permutations σ, of the terms $Lk(\mathbb{L};\sigma)$signature(σ). (The signature of a permutation σ of (n+1) elements with (s+1) circles in its graph is $(-1)^{n+1+s+1} = (-1)^{n+s}$.) This proves 3.6.4.

❑

Similarly,

3.6.5 $\alpha(N\backslash J\,;\, r — [r=0];\, s\,O\,;\, (-1)^{n+\#J+s}) = \det(E(\mathbb{L}_{N\backslash J}))$

❑

Now, according to 3.6.4 and 3.6.5, Assertion 3.6.2 will be the consequence of Assertions 3.6.6 and 3.6.7.

3.6.6
For any non empty subset J *of* N,

$$v(J) = \omega p_0 \alpha(N\backslash J\,;\, r —\,;\, s\,O\,;\, (-1)^{n+\#J+s}\, r!\, \omega^r)$$
$$+\, q_0 \alpha(N\backslash J\,;\, r — [r\geq 1];\, s\,O\,;\, (-1)^{n+\#J+s}\, r!\, \omega^r)$$
$$v(\emptyset) = \omega p_0 \alpha(N\,;\, r — [r\geq 1];\, s\,O\,;\, (-1)^{n+s}\, r!\, \omega^r)$$
$$+\, q_0 \alpha(N\,;\, r — [r\geq 2];\, s\,O\,;\, (-1)^{n+s}\, r!\, \omega^r)$$

3.6.7 *For any subset* J *of* N,

$$u(J) - q_0 \det(E(\mathbb{L}_{N\backslash J})) =$$
$$\omega p_0 \det(E(\mathbb{L}_{N\backslash J})) + (q_0+p_0\omega)\alpha(N\backslash J\,;\, r — [r\geq 1];\, s\,O\,;\, (-1)^{n+\#J+s}\, r!\, \omega^r)$$

❑

PROOF OF 3.6.6:
This is nothing more than expanding $\det(E(\mathbb{L}_{N\cup\{0\}\backslash I}))$; the first term in the right-hand side of 3.6.6 corresponds to the permutations of $N\cup\{0\}\backslash I$ which fix 0, the other one corresponds to the permutations which do not fix 0. When J is empty, the additional restrictions on r come from the fact that I cannot be empty in the expression of v(J) given in 3.6.1.

❑

PROOF OF 3.6.7
Let w(J) be defined by:

$$w(J) = \frac{u(J)}{q_0 + p_0\omega}$$
$$= \sum_{\{I\,/\,J\subset I\subset N\}} \Big((-1)^{\#I+\#J} \alpha(I\backslash J\,;\, i —\,;\, (i+1)!\omega^i)\, \det(E(\tilde{\mathbb{L}}_{N\backslash I})) \Big)$$

We prove the following equality implying 3.6.7:

3.6.8 $w(J) = \alpha(N\backslash J\,;\, r —\,;\, s\,O\,;\, (-1)^{n+\#J+s}\, r!\, \omega^r)$

We compute w(J) by cutting the permutation graphs $G(\rho)$ involved in the $\det(E(\tilde{\mathbb{L}}_{N\backslash I}))$ (r-i) times, leaving s circles uncut. The permutation ρ equips the

set Γ of the (r-i) lines, created by the cuts, with a permutation σ mapping a line to the line right after it on the cycle of ρ where it comes from.
So we get:

3.6.9

$$w(J) = \alpha(N\backslash J\ ;\ r\ —\ ;\ s\ O\ ;\ c(r,s))$$

with

$$c(r,s) = \omega^r \sum_{i=0}^{r} (i+1)! \sum_{\{\Gamma / \Gamma \subset \{1,\dots, r\},\ \#\Gamma = r-i\}} \sum_{\sigma\in\sigma_\Gamma} \mathrm{signature}(\sigma)(-1)^{n+\#J+s+r-i}$$

where σ_Γ denotes the set of permutations of Γ.
Using

3.6.10 $$\sum_{\sigma\in\sigma_\Gamma} \mathrm{signature}(\sigma) = \begin{cases} 0 & \text{if } \#\Gamma\geq 2 \\ 1 & \text{if } \#\Gamma\leq 1 \end{cases}$$

gives

$$c(r,s) = (-1)^{n+\#J+s}\ r!\ \omega^r$$

This proves 3.6.8,

❑

and completes the proof of 3.2.11.

❑ ❑

Proposition 3.1.1 is now completely proved. Thus we know that setting:

$$\lambda(\chi_{S^3}(\mathbb{L})) = \mathbb{F}_{S^3}(\mathbb{L}) \quad \textit{for any surgery presentation } \mathbb{L} \textit{ in } S^3$$

provides a consistent definition of an invariant λ of closed oriented 3-manifolds. This invariant will be denoted by λ from now on.

Chapter 4

The formula for surgeries starting from rational homology spheres

§4.1 Introduction

§4.2 to §4.5 prove the surgery formula T2 satisfied by λ, that is:

PROPOSITION T2:
For any rational homology sphere M, *and for any surgery presentation* $\mathbb{H}$ *in* M, *the surgery formula* $F(M,\mathbb{H})$ *is satisfied:*

$$(\ \mathbf{F(M,\mathbb{H})}\ ) \qquad \lambda(\chi_M(\mathbb{H})) = \frac{|H_1(\chi_M(\mathbb{H}))|}{|H_1(M)|}\lambda(M) + \mathbb{F}_M(\mathbb{H})$$

❑

Section 4.7 relates the one-component surgery formula to the Walker formula, recalled in §4.6. This relationship implies that λ satisfies Property T5.0:

PROPERTY T5.0:
If M *is a rational homology sphere, and if* λ_W *denotes the Walker invariant as described in* [W]:

$$\lambda(M) = \frac{|H_1(M)|}{2}\lambda_W(M)$$

❑

§4.2 Sketch of the proof of Proposition T2

By definition, Proposition T2 holds for all surgery presentations in S^3. Our approach to the proof of Proposition T2 will be to exhibit sufficiently many equivalences between (the validity of) surgery formulae $F(M,\mathbb{H})$ (for different $(M,\mathbb{H})$) to prove that all of these formulae are equivalent, and hence all true since the formulae $F(S^3,\mathbb{H})$ are true.

NOTATION **4.2.1**:
$<.,.>$ denotes the intersection form on the oriented boundaries of the tubular neighborhoods of the knots. If m and ℓ are an oriented meridian and an oriented parallel of a knot, respectively,

$$<m,\ell> = 1$$

LEMMA **4.2.2**:
HYPOTHESES (and notation) OF LEMMA **4.2.2:**
Let S be a rational homology sphere, and let $\mathbb{L}_A = (K_i,\mu_i)_{i\in A}$ denote a surgery presentation in S.
(A is the set of component indices of $\mathbb{L}_A$.)
Let N denote a subset of A such that $\chi_S(\mathbb{L}_N)$ is a rational homology sphere.
Let M denote $\chi_S(\mathbb{L}_N)$, let $\mathbb{H}_{A\setminus N}$ (respectively $H_{A\setminus N}$) denote the surgery presentation $\mathbb{L}_{A\setminus N}$ (respectively its underlying link), when viewed in $M = \chi_S(\mathbb{L}_N)$. ($\chi_M(\mathbb{H}_{A\setminus N}) = \chi_S(\mathbb{L}_A)$)
Let $(*\mathbb{L}_A,N)$ denote the following equality:
$(*\mathbb{L}_A,N)$

$$\left(\prod_{i\in A} q_i\right) \sum_{\{I\,/\,I\neq\emptyset\,,\,I\subset A\}} \det(E(\mathbb{L}_{A\setminus I}))\,\zeta_A(L_I)$$

$$=$$

$$\det(E(\mathbb{H}_{A\setminus N}))\left(\prod_{i\in A} q_i\right) \sum_{\{I\,/\,I\neq\emptyset\,,\,I\subset N\}} \det(E(\mathbb{L}_{N\setminus I}))\,\zeta_N(L_I)$$

$$+\ \mathrm{sign}(\mathbb{L}_N)\left(\prod_{i\in A\setminus N} q_i\right) \sum_{\{I\,/\,I\neq\emptyset\,,\,I\subset A\setminus N\}} \det(E(\mathbb{H}_{(A\setminus N)\setminus I}))\,\zeta_{A\setminus N}(H_I)$$

$$+\ \frac{|H_1(S)|}{24}\left(\prod_{i\in A} q_i\right)\sum_{i\in N} \frac{\det(E(\mathbb{L}_{A\setminus\{i\}})) - \det(E(\mathbb{H}_{A\setminus N}))\det(E(\mathbb{L}_{N\setminus\{i\}}))}{q_i^2}$$

$$+\ \frac{|H_1(S)|\det(E(\mathbb{L}_A))}{24}\left(\prod_{i\in A} q_i\right)\sum_{i\in A\setminus N}\left(\frac{Lk_S(\mu_i,K_i) - Lk_M(\mu_i,K_i)}{q_i}\right)$$

(As usual, the positive intersection number $<m_i,\mu_i>$ of an oriented meridian m_i of K_i and of the characteristic curve μ_i of the surgery performed on K_i is denoted by q_i.)
Assume that $F(S,\mathbb{L}_N)$ is true.

STATEMENT OF LEMMA **4.2.2:**
Under the hypotheses above, if $(*\mathbb{L}_A,N)$ *is satisfied, then* $F(S,\mathbb{L}_A)$ *and* $F(M,\mathbb{H}_{A\backslash N})$ *are equivalent; and, conversely, if* $F(S,\mathbb{L}_A)$ *and* $F(M,\mathbb{H}_{A\backslash N})$ *are both true, then* $(*\mathbb{L}_A,N)$ *is satisfied.*

❑

LEMMA **4.2.3**:
Under the hypotheses of Lemma 4.2.2, *if* $\mathbb{H}_{A\backslash N}$ *has only one component* K, *and if the linking number (in* M *) of* K *with any core of the surgery performed on* $\mathbb{L}_N$ *is nonzero, then* $(*\mathbb{L}_A,N)$ *is satisfied.*

❑

So, according to Lemma 4.2.2, under the hypotheses of Lemma 4.2.3, $F(S,\mathbb{L}_A)$ and $F(M,\mathbb{H}_{A\backslash N})$ are equivalent.
These two lemmas, which will be proved later, show in particular:

4.2.4 *The surgery formula* $F(M,\mathbb{H})$ *is true for any rational homology sphere* M *and for any one-component surgery presentation* $\mathbb{H}$.
Indeed, it suffices to present M with a surgery presentation $\mathbb{L}_N$ in S^3, and to isotope the underlying knot K of $\mathbb{H}$ in M so that it algebraically links the cores of the surgery performed on $\mathbb{L}_N$, in order to satisfy all the hypotheses of the lemmas with $S = S^3$ and $\mathbb{H}_{A\backslash N} = \mathbb{H}$.

❑

In fact, §4.5 will show that 4.2.2 and 4.2.3 are also sufficient to prove:

LEMMA **4.2.5**:
Let M *be a rational homology sphere and let* H *be a link in* M *such that any two components of* H *have a nonzero linking number. Then* $F(M, \mathbb{H})$ *is true for any surgery presentation* $\mathbb{H}$ *in* M *with underlying link* H.

❑

PROOF OF PROPOSITION T2
(Assuming Lemmas 4.2.2, 4.2.3, 4.2.5 which will be proved in §4.3, §4.4 and §4.5, respectively.)

Let $\mathbb{H}$ be a surgery presentation in a rational homology sphere M. It can be transformed into a surgery presentation $\mathbb{H}'$ with an underlying link H' satisfying the hypothesis of Lemma 4.2.5, by adding first to $\mathbb{H}$ a trivial component K_0 in M with its meridian as characteristic curve μ_0, and by performing next an

ω-twist with boundary K_0 on $\mathbb{H} \cup (K_0, \mu_0)$. (Referring strictly to Definition 1.6.4, $\mathbb{H}$ is the result of a $(-\omega)$-twist on $\mathbb{H}'$.)
(Choose for example K_0 so that its linking number with any component of H is 1; then, an ω-twist adds ω to the linking number of any two initial components of H (see 3.1.3); thus it is easy to choose ω so that any two components of H' algebraically link each other.)
According to the definition of an ω-twist (1.6.4), $\chi_M(\mathbb{H})$ and $\chi_M(\mathbb{H}')$ are homeomorphic, and according to Proposition 3.1.1, $\mathbb{F}_M(\mathbb{H})$ and $\mathbb{F}_M(\mathbb{H}')$ are equal. So $F(M,\mathbb{H})$ is equivalent to $F(M,\mathbb{H}')$ which is true according to Lemma 4.2.5, and we are done.

❑

§4.3 Proof of Lemma 4.2.2

We assume that the hypotheses of Lemma 4.2.2 are satisfied.
Since $\chi_S(\mathbb{L}_A)$ and $\chi_M(\mathbb{H}_{A\setminus N})$ are homeomorphic and since, by hypothesis, $F(S,\mathbb{L}_N)$ is true, to prove Lemma 4.2.2 it suffices to show that $(*\mathbb{L}_A,N)$ is equivalent to:

$$\textbf{4.3.1} \qquad \frac{|H_1(\chi_S(\mathbb{L}_A))|}{|H_1(S)|}\lambda(S) + \mathbb{F}_S(\mathbb{L}_A)$$
$$= \frac{|H_1(\chi_M(\mathbb{H}_{A\setminus N}))|}{|H_1(M)|}\left(\frac{|H_1(M)|}{|H_1(S)|}\lambda(S) + \mathbb{F}_S(\mathbb{L}_N)\right) + \mathbb{F}_M(\mathbb{H}_{A\setminus N})$$

Equality 4.3.1 is equivalent to 4.3.2:

$$\textbf{4.3.2} \qquad \mathbb{F}_S(\mathbb{L}_A) = \frac{|H_1(\chi_M(\mathbb{H}_{A\setminus N}))|}{|H_1(M)|}\mathbb{F}_S(\mathbb{L}_N) + \mathbb{F}_M(\mathbb{H}_{A\setminus N})$$

Let us use 1.3.5, Definition 1.7.3 and Remark 1.7.4 to decompose $\mathbb{F}$ in the following way:
(We write down the decomposition for $\mathbb{F}_S(\mathbb{L}_A)$ and we use similar decompositions for $\mathbb{F}_S(\mathbb{L}_N)$ and $\mathbb{F}_M(\mathbb{H}_{A\setminus N})$.)

$$\textbf{4.3.3} \qquad \mathbb{F}_S(\mathbb{L}_A) = (a+b+c+d+e)(\mathbb{L}_A)$$

with

$$a(\mathbb{L}_A) = \text{sign}(\mathbb{L}_A)\left(\prod_{i\in A} q_i\right)\sum_{\{I \,/\, I\neq\emptyset\,,\, I\subset A\}} \det(E(\mathbb{L}_{A\setminus I}))\,\zeta_A(L_I)$$

$$b(\mathbb{L}_A) = -\,\text{sign}(\mathbb{L}_A)\sum_{i\in A}\frac{|H_1(\chi(\mathbb{L}_{A\setminus\{i\}}))|}{24\, q_i\, \text{sign}(\mathbb{L}_{A\setminus\{i\}})}$$

$$c(\mathbb{L}_A) = \frac{|H_1(\chi(\mathbb{L}_A))|\ \text{signature}\,(\,E(\mathbb{L}_A)\,)}{8}$$

$$d(\mathbb{L}_A) = |H_1(\chi(\mathbb{L}_A))| \sum_{i\in A} \frac{s(<\mu_i, \ell_i>, q_i)}{2}$$

$$e(\mathbb{L}_A) = -\, |H_1(\chi(\mathbb{L}_A))| \sum_{i\in A} \frac{p_i}{24 q_i}$$

where $<\mu_i, \ell_i>$ is the intersection number of μ_i and an oriented parallel ℓ_i of K_i.

This decomposition makes clear that the d-terms involving Dedekind sums cancel out in 4.3.2.
The terms of $e(\mathbb{L}_A)$ with indices i in N and $e(\mathbb{L}_N)$ obviously cancel each other.

The following sublemma shows that the c-terms involving signatures cancel out in 4.3.2.

SUBLEMMA **4.3.4**:
The quadratic form represented by $E(\mathbb{L}_A)$ *is the orthogonal sum of the quadratic forms represented by* $E(\mathbb{L}_N)$ *and* $E(\mathbb{H}_{A\setminus N})$ *(over* $\mathbb{R}$*).*

❑

Sublemma 4.3.4 also yields:

4.3.5 $\quad \operatorname{sign}(\mathbb{L}_A) = \operatorname{sign}(\mathbb{L}_N)\, \operatorname{sign}(\mathbb{H}_{A\setminus N})$

and, similarly, for $i \in A\setminus N$,

$$\operatorname{sign}(\mathbb{L}_{A\setminus\{i\}}) = \operatorname{sign}(\mathbb{L}_N)\, \operatorname{sign}(\mathbb{H}_{(A\setminus\{i\})\setminus N})$$

So,

$$\frac{\operatorname{sign}(\mathbb{L}_A)}{\operatorname{sign}(\mathbb{L}_{A\setminus\{i\}})} = \frac{\operatorname{sign}(\mathbb{H}_{A\setminus N})}{\operatorname{sign}(\mathbb{H}_{(A\setminus\{i\})\setminus N})}$$

This proves that the terms of $b(\mathbb{L}_A)$ with indices i in $A\setminus N$ and $b(\mathbb{H}_{A\setminus N})$ cancel each other.

All these cancellations prove Lemma 4.2.2 except for Sublemma 4.3.4.

❑

PROOF OF SUBLEMMA 4.3.4
Since $E(\mathbb{L}_N)$ and $E(\mathbb{L}_N)^{-1}$ represent the same quadratic form, it suffices to prove that:

4.3.6 $E(\mathbb{L}_A)$ and $\begin{pmatrix} E(\mathbb{H}_{A\setminus N}) & 0 \\ 0 & E(\mathbb{L}_N)^{-1} \end{pmatrix}$ represent the same quadratic form,

to prove:
4.3.4 The quadratic form represented by $E(\mathbb{L}_A)$ is the orthogonal sum of the quadratic forms represented by $E(\mathbb{L}_N)$ and $E(\mathbb{H}_{A\setminus N})$.

The matrices $E(\mathbb{L}_A) = \begin{pmatrix} E(\mathbb{L}_{A\setminus N}) & X \\ {}^tX & E(\mathbb{L}_N) \end{pmatrix}$ and

$$\begin{pmatrix} E(\mathbb{L}_{A\setminus N}) - XE(\mathbb{L}_N)^{-1}({}^tX) & 0 \\ 0 & E(\mathbb{L}_N)^{-1} \end{pmatrix}$$

$$\left(= \begin{pmatrix} Id_{A\setminus N} & -XE(\mathbb{L}_N)^{-1} \\ 0 & E(\mathbb{L}_N)^{-1} \end{pmatrix} \begin{pmatrix} E(\mathbb{L}_{A\setminus N}) & X \\ {}^tX & E(\mathbb{L}_N) \end{pmatrix} \begin{pmatrix} Id_{A\setminus N} & 0 \\ -E(\mathbb{L}_N)^{-1}({}^tX) & E(\mathbb{L}_N)^{-1} \end{pmatrix} \right)$$

represent the same quadratic form. So, it suffices now to prove:

$$\textbf{4.3.7} \qquad E(\mathbb{L}_{A\setminus N}) - XE(\mathbb{L}_N)^{-1}({}^tX) = E(\mathbb{H}_{A\setminus N})$$

Equality 4.3.7 will be shown by viewing all the matrices involved as matrices of system transformations of $H_1(S\setminus L_A;\mathbb{Q})$. ($L_A$ is the underlying link of $\mathbb{L}_A$.) The systems will be written as row vectors. For example, m_A denotes the basis $(m_a)_{a\in A}$ of the meridians of L_A in S, written as a row; and μ_A denotes the system $(\mu_a)_{a\in A}$ of the characteristic curves of $\mathbb{L}_A$. With this notation:

$$\left(\left(\frac{1}{<m,\mu>}\mu\right)_{A\setminus N}, \left(\frac{1}{<m,\mu>}\mu\right)_N\right) = (m_{A\setminus N}, m_N) \begin{pmatrix} E(\mathbb{L}_{A\setminus N}) & X \\ {}^tX & E(\mathbb{L}_N) \end{pmatrix}$$

So,

$$\begin{cases} \left(\frac{1}{<m,\mu>}\mu\right)_{A\setminus N} = m_{A\setminus N}E(\mathbb{L}_{A\setminus N}) + m_N{}^tX \\ \left(\frac{1}{<m,\mu>}\mu\right)_N - m_{A\setminus N}X = m_NE(\mathbb{L}_N) \end{cases}$$

and

$$\left(\frac{1}{<m,\mu>}\mu\right)_{A\setminus N} = m_{A\setminus N}E(\mathbb{L}_{A\setminus N}) + \left(\left(\frac{1}{<m,\mu>}\mu\right)_N - m_{A\setminus N}X\right)E(\mathbb{L}_N)^{-1}({}^tX)$$

$$= m_{A\setminus N}\left(E(\mathbb{L}_{A\setminus N}) - XE(\mathbb{L}_N)^{-1}({}^tX)\right) + \left(\frac{1}{<m,\mu>}\mu\right)_N E(\mathbb{L}_N)^{-1}({}^tX)$$

Thus, in $H_1(M\setminus L_{A\setminus N};\mathbb{Q})$,

$$\left(\frac{1}{<m,\mu>}\mu\right)_{A\setminus N} = m_{A\setminus N}\left(E(\mathbb{L}_{A\setminus N}) - XE(\mathbb{L}_N)^{-1}({}^tX)\right)$$

This proves 4.3.7, and completes the proof of Sublemma 4.3.4.

❑

§4.4 Proof of Lemma 4.2.3

Let us rewrite Lemma 4.2.3 in order to fix notations:

LEMMA **4.2.3**:

Let S be a rational homology sphere and let

$$\mathbb{L} = \mathbb{L}_{N\cup\{0\}} = (K_i\,;\mu_i)_{i\in N\cup\{0\}}$$

be a surgery presentation in S, such that $M = \chi_S(\mathbb{L}_N)$ is a rational homology sphere and $F(S,\mathbb{L}_N)$ is true. Let $\mathbb{H}$ denote the surgery presentation (K_0,μ_0) in M. Let $C_i \subset M$ denote the core of the surgery performed on K_i for $i \in N = \{1, ..., n\}$.
Assume that $Lk_M(K_0,C_i)$ *is different from* 0, *for any* i *of* N. *Then, Equality* $(*\mathbb{L}_{N\cup\{0\}},N)$ *is satisfied.*

SPLITTING THE PROOF OF LEMMA 4.2.3

The equality $(*\mathbb{L}_{N\cup\{0\}},N)$ to be proved is:
$(*\mathbb{L}_{N\cup\{0\}},N)$

$$\left(\prod_{i\in N\cup\{0\}} q_i\right) \sum_{\{I / I\neq\emptyset, I\subset N\cup\{0\}\}} \det(E(\mathbb{L}_{N\cup\{0\}\setminus I}))\, \zeta_{N\cup\{0\}}(L_I)$$
$$= \det(E(\mathbb{H})) \left(\prod_{i\in N\cup\{0\}} q_i\right) \sum_{\{I / I\neq\emptyset, I\subset N\}} \det(E(\mathbb{L}_{N\setminus I}))\, \zeta_N(L_I)$$
$$+ \operatorname{sign}(\mathbb{L}_N)\, q_0\, \zeta(K_0\subset M)$$
$$+ \frac{|H_1(S)|}{24} \left(\prod_{i\in N\cup\{0\}} q_i\right) \sum_{i\in N} \frac{\det(E(\mathbb{L}_{N\cup\{0\}\setminus\{i\}})) - \det(E(\mathbb{H}))\det(E(\mathbb{L}_{N\setminus\{i\}}))}{q_i^2}$$
$$+ \frac{|H_1(S)|\det(E(\mathbb{L}_{N\cup\{0\}}))}{24} \left(\prod_{i\in N\cup\{0\}} q_i\right) \left(\frac{Lk_S(\mu_0,K_0) - Lk_M(\mu_0,K_0)}{q_0}\right)$$

It will be the direct consequence of the three following lemmas:

LEMMA **4.4.1**:

$$\left(\prod_{i\in N\cup\{0\}} q_i\right) \sum_{\{I / I\neq\emptyset, I\subset N\cup\{0\}\}} \det(E(\mathbb{L}_{N\cup\{0\}\setminus I}))\, \zeta_{N\cup\{0\}}(L_I)$$
$$= \det(E(\mathbb{H})) \left(\prod_{i\in N\cup\{0\}} q_i\right) \sum_{\{I / I\neq\emptyset, I\subset N\}} \det(E(\mathbb{L}_{N\setminus I}))\, \zeta_{N\cup\{0\}}(L_I)$$
$$+ \operatorname{sign}(\mathbb{L}_N)\, q_0\, \zeta_{N\cup\{0\}}(K_0\subset M)$$

❑

LEMMA **4.4.2**:

$$\operatorname{sign}(\mathbb{L}_N)\, q_0\, \zeta_{N\cup\{0\}}(K_0\subset M) = \operatorname{sign}(\mathbb{L}_N)\, q_0\, \zeta_{\{0\}}(K_0\subset M)$$
$$+ \frac{|H_1(S)|}{24} \left(\prod_{i\in N\cup\{0\}} q_i\right) \sum_{i\in N} \frac{\det(E(\mathbb{L}_{N\cup\{0\}\setminus\{i\}})) - \det(E(\mathbb{H}))\det(E(\mathbb{L}_{N\setminus\{i\}}))}{q_i^2}$$

❑

LEMMA **4.4.3**:

$$\det(E(\mathbb{H}))\left(\prod_{i\in N\cup\{0\}} q_i\right)\sum_{\{I\,/\,I\neq\emptyset\,,\,I\subset N\}}\det(E(\mathbb{L}_{N\setminus I}))\,\zeta_{N\cup\{0\}}(L_I)$$

$$= \det(E(\mathbb{H}))\left(\prod_{i\in N\cup\{0\}} q_i\right)\sum_{\{I\,/\,I\neq\emptyset\,,\,I\subset N\}}\det(E(\mathbb{L}_{N\setminus I}))\,\zeta_{N}(L_I)$$

$$+\ \frac{|H_1(S)|\det(E(\mathbb{L}_{N\cup\{0\}}))}{24}\left(\prod_{i\in N\cup\{0\}} q_i\right)\left(\frac{Lk_S(\mu_0,K_0) - Lk_M(\mu_0,K_0)}{q_0}\right)$$

❑

PROOF OF LEMMA 4.4.3

Let us first observe that:

$$\textbf{4.4.4} \qquad \det(E(\mathbb{H})) = \frac{Lk_M(\mu_0,K_0)}{q_0} = \frac{\det(E(\mathbb{L}_{N\cup\{0\}}))}{\det(E(\mathbb{L}_N))}$$

Lemma 4.4.3 will then follow from Equality 4.4.5:

4.4.5

$$\sum_{\{I\,/\,I\neq\emptyset\,,\,I\subset N\}}(-1)^{\#I+1}\det(E(\mathbb{L}_{N\setminus I}))\,Lk_c(L_{I\cup\{0\}})$$

$$= \det(E(\mathbb{L}_N))\left(\frac{Lk_S(\mu_0,K_0) - Lk_M(\mu_0,K_0)}{q_0}\right)$$

Equality 4.4.5 is true because both of its sides are equal to:

$$\left(\frac{Lk_S(\mu_0,K_0)}{q_0}\det(E(\mathbb{L}_N)) - \det(E(\mathbb{L}_{N\cup\{0\}}))\right)$$

(For the left-hand side, expand $(-\det(E(\mathbb{L}_{N\cup\{0\}})))$ as the usual sum over all permutations of $N\cup\{0\}$; the left-hand side of 4.4.5 then corresponds to the permutations which do not fix 0.)

❑

The proofs of Lemmas 4.4.1 and 4.4.2 require more notation and a combinatorial lemma.

NOTATION **4.4.6**:

- $\ell_{ij} = \dfrac{Lk_S(\mu_j,K_i)}{q_j}$
- m_i denotes the oriented meridian of K_i in S.
- φ denotes the function $Lk_M(K_0,.)$ on $H_1(M\setminus K_0;\mathbb{Q})$.
- L denotes the underlying link of $\mathbb{L}$ in S, and C denotes the link $C = K_0, C_1, C_2, \ldots, C_n$ in M.
- For $i = 1, \ldots, n$, the knot C_i is oriented so that an oriented parallel λ_i of C_i satisfies:

$$<\mu_i,\lambda_i> = 1$$

Since

$$<\mu_i,m_i> = - q_i$$

m_i is homologous to $(- q_iC_i)$ in $M\backslash K_0$. This gives the relation:

$$\textbf{4.4.7} \qquad \varphi(m_i) = - q_i\varphi(C_i)$$

• For any subset I of $N\cup\{0\}$, we define the coefficient B(I) by:

$$\textbf{4.4.8} \qquad B(I) = (-1)^{n+1+\#I} \sum_{g \in \mathcal{F}(N\cup\{0\}\backslash I\,,\ \to I)} \prod_{i\in N\cup\{0\}\backslash I} \ell_{ig(i)}\varphi(m_{g(i)})$$

LEMMA **4.4.9**:

For any subset I of $N\cup\{0\}$,

$$B(I) = \begin{cases} \prod_{i\in N\cup\{0\}\backslash I} \varphi(m_i)\det(E(\mathbb{L}_{N\cup\{0\}\backslash I})) & \text{if } 0 \in I \\ \prod_{i\in N\cup\{0\}\backslash I} \varphi(m_i)\Big(\det(E(\mathbb{L}_{N\cup\{0\}\backslash I})) - \det(E(\mathbb{H}))\det(E(\mathbb{L}_{N\backslash I}))\Big) & \text{if } 0 \notin I \end{cases}$$

❑

PROOF OF LEMMA 4.4.1 ASSUMING LEMMA 4.4.9

The equality to be proved can be written as:

4.4.10

$$\zeta_{N\cup\{0\}}(K_0\subset M) = \mathrm{sign}(\mathbb{L}_N)\left(\prod_{i\in N} q_i\right) \sum_{\{I\,/\,I\neq\emptyset\,,\,I\subset N\cup\{0\}\}} \det(E(\mathbb{L}_{N\cup\{0\}\backslash I}))\,\zeta_{N\cup\{0\}}(L_I)$$
$$- \det(E(\mathbb{H}))\,\mathrm{sign}(\mathbb{L}_N)\left(\prod_{i\in N} q_i\right) \sum_{\{I\,/\,I\neq\emptyset\,,\,I\subset N\}} \det(E(\mathbb{L}_{N\backslash I}))\,\zeta_{N\cup\{0\}}(L_I)$$

Let T be the multiindex (n+1, 0, ..., 0).

On the one hand, according to Definition 2.4.6 (see also Example 2.4.8 part 2), the only nonzero contribution to the expression of $\mathcal{D}_T(C)$ given by Proposition 2.5.2 occurs for $I=\{0\}$ and therefore

$$\mathcal{D}_T(C) = (-1)^n\left(\prod_{i=1}^{n}\varphi(C_i)\right)\zeta_{N\cup\{0\}}(K_0\subset M)$$

On the other hand, $\mathcal{D}_T(C)$ is the coefficient of $(m_0)^{n+1}$ in $\mathcal{D}(C)(m_0, \mu_1, ..., \mu_n)$ (here m_0 and the μ_i's denote variables). So, according to 2.3.2, $\mathcal{D}_T(C)$ must be the coefficient of $(m_0)^{n+1}$ in $\mathrm{sign}(\mathbb{L}_N)\mathcal{D}(L)(m_0, \varphi(m_1)m_0, ..., \varphi(m_n)m_0)$. Thus,

$$\mathrm{sign}(\mathbb{L}_N)\mathcal{D}_T(C) = \sum_{\{R\,/\,R \text{ is a multiindex, } |R| = n+1,\ S(R) = N\cup\{0\}\}} \left(\prod_{i=0}^{n} \varphi(m_i)^{r_i}\right) \mathcal{D}_R(L)$$

Applying again Proposition 2.5.2 to compute the $\mathcal{D}_R(L)$ yields:

$$\mathrm{sign}(\mathbb{L}_N)\mathcal{D}_T(C)$$

$$= \sum_{\substack{\{(I,f)\,/\,I\subset N\cup\{0\},\\ f\in \mathcal{F}(N\cup\{0\}\setminus I,\ \to I)\}}} \left((-1)^{n+1\,-\,\#I} \left(\prod_{i\in N\cup\{0\}\setminus I} \ell_{if(i)}\varphi(m_{f(i)})\right)\left(\prod_{i\in I}\varphi(m_i)\right) \zeta_{N\cup\{0\}}(L_I)\right)$$

$$= \sum_{\{I\subset N\cup\{0\}\}} \left(B(I) \left(\prod_{i\in I}\varphi(m_i)\right) \zeta_{N\cup\{0\}}(L_I)\right)$$

Comparing these two expressions of $\mathcal{D}_T(C)$ with the help of 4.4.7 and 4.4.9 proves 4.4.10.

❑

PROOF OF LEMMA 4.4.2 ASSUMING LEMMA 4.4.9

The equality to be proved is:

4.4.2

$$\mathrm{sign}(\mathbb{L}_N)\, q_0\, \zeta_{N\cup\{0\}}(K_0\subset M) = \mathrm{sign}(\mathbb{L}_N)\, q_0\, \zeta_{\{0\}}(K_0\subset M)$$
$$+ \frac{|H_1(S)|}{24}\left(\prod_{i\in N\cup\{0\}} q_i\right)\sum_{i\in N}\frac{\det(E(\mathbb{L}_{N\cup\{0\}\setminus\{i\}})) - \det(E(\mathbb{H}))\det(E(\mathbb{L}_{N\setminus\{i\}}))}{q_i^2}$$

where, according to 1.7.1 and 1.3.5,

$$\zeta_{N\cup\{0\}}(K_0\subset M) - \zeta_{\{0\}}(K_0\subset M)$$
$$= \mathrm{sign}(\mathbb{L}_N)\,\frac{|H_1(S)|\det(E(\mathbb{L}_N))}{24}\left(\prod_{i\in N} q_i\right)\sum_{i\in N}\varphi(C_i)^2$$

and, according to 4.4.7,

$$q_i^2\varphi(C_i)^2 = \varphi(m_i)^2$$

So, it suffices to prove, for $i \in N$:

$$\det(E(\mathbb{L}_N))\,\varphi(m_i)^2 = \det(E(\mathbb{L}_{N\cup\{0\}\setminus\{i\}})) - \det(E(\mathbb{H}))\det(E(\mathbb{L}_{N\setminus\{i\}}))$$

According to 4.4.9, this is equivalent to

$$\textbf{4.4.11} \qquad \varphi(m_0)\, B(\{0\})\, \varphi(m_i)^2 = \varphi(m_i)\, B(\{i\})$$

Observe that, according to 2.4.8 and 4.4.8, for $i\in N\cup\{0\}$,

$$\textbf{4.4.12} \qquad \frac{B(\{i\})}{\varphi(m_i)} = (-1)^n \sum_R \mathrm{Lk}(L;R) \prod_{j\in N\cup\{0\}} \varphi(m_j)^{r_j}$$

where the sum runs over all multiindices R with support $N\cup\{0\}$ and modulus $(n-1)$. Thus $\frac{B(\{i\})}{\varphi(m_i)}$ is independent of i. Since $\varphi(m_0)$ equals 1, this proves 4.4.11, and 4.4.2.

❑

PROOF OF LEMMA 4.4.9

Let I be a subset of $N\cup\{0\}$, the equality to be proved is:

4.4.9

$$(-1)^{n+1+\#I} \sum_{g\in\mathfrak{F}(N\cup\{0\}\setminus I\,,\,\to I)} \prod_{i\in N\cup\{0\}\setminus I} \ell_{ig(i)}\varphi(m_{g(i)}) =$$

$$\begin{cases} \prod_{i\in N\cup\{0\}\setminus I} \varphi(m_i)\det(E(\mathbb{L}_{N\cup\{0\}\setminus I})) & \text{if } 0\in I \\ \prod_{i\in N\cup\{0\}\setminus I} \varphi(m_i)\big(\det(E(\mathbb{L}_{N\cup\{0\}\setminus I})) - \det(E(\mathbb{H}))\det(E(\mathbb{L}_{N\setminus I}))\big) & \text{if } 0\notin I \end{cases}$$

A function g of $\mathfrak{F}(N\cup\{0\}\setminus I\,,\,\to I)$ decomposes into a function from a set $P\subset N\cup\{0\}\setminus I$ to I and a function g of $\mathfrak{F}(N\cup\{0\}\setminus(I\cup P)\,,\,\to P)$. This decomposition is unique. (P is the inverse image of I under g.) Hence,

4.4.13

$$\sum_{g\in\mathfrak{F}(N\cup\{0\}\setminus I\,,\,\to I)} \prod_{i\in N\cup\{0\}\setminus I} \ell_{ig(i)}\varphi(m_{g(i)}) =$$

$$\sum_{P\subset(N\cup\{0\}\setminus I)} \left(\sum_{g\in\mathfrak{F}(N\cup\{0\}\setminus(I\cup P)\,,\,\to P)} \prod_{i\in N\cup\{0\}\setminus(I\cup P)} \ell_{ig(i)}\varphi(m_{g(i)})\right)\left(\sum_{g\in\mathfrak{F}(P,I)} \prod_{i\in P} \ell_{ig(i)}\varphi(m_{g(i)})\right)$$

where $\mathfrak{F}(P,I)$ denotes the set of functions from P to I.

$$\sum_{g\in\mathfrak{F}(P,I)} \prod_{p\in P} \big(\ell_{pg(p)}\varphi(m_{g(p)})\big) = \prod_{p\in P} \sum_{i\in I} \ell_{pi}\varphi(m_i)$$

Note that:

$$\begin{cases} \sum_{i\in N\cup\{0\}} \ell_{pi}\varphi(m_i) = \frac{\varphi(\mu_p)}{q_p} = 0 & \forall p\in N \\ \sum_{i\in N} \ell_{0i}\varphi(m_i) = \frac{\varphi(\mathfrak{L}_0)}{\varphi(m_0)}\varphi(m_0) \end{cases}$$

where $\mathcal{L}_0$ is the element of $H_1(\partial T(K_0);\mathbb{Q})$, which is zero in $H_1(S\backslash K_0;\mathbb{Q})$, and such that $<m_0, \mathcal{L}_0> = 1$.
Thus, setting

$$\tilde{\ell}_{ij} = \begin{cases} \ell_{ij} & \text{if } (i,j) \neq (0,0) \\ -\dfrac{\varphi(\mathcal{L}_0)}{\varphi(m_0)} & \text{if } (i,j) = (0,0) \end{cases}$$

gives:

4.4.14 $\forall p \in N \cup \{0\}$,

$$\sum_{i \in N \cup \{0\}} \tilde{\ell}_{pi} \varphi(m_i) = 0$$

So:

$$\prod_{p \in P} \sum_{i \in I} \ell_{pi} \varphi(m_i) = (-1)^{\#P} \prod_{p \in P} \sum_{i \in N \cup \{0\} \backslash I} \tilde{\ell}_{pi} \varphi(m_i)$$

$$= (-1)^{\#P} \sum_{g \in \mathcal{F}(P, N \cup \{0\} \backslash I)} \prod_{p \in P} \tilde{\ell}_{pg(p)} \varphi(m_{g(p)})$$

Therefore 4.4.13 becomes:

$$\sum_{g \in \mathcal{F}(N \cup \{0\} \backslash I, \rightarrow I)} \prod_{i \in N \cup \{0\} \backslash I} \ell_{ig(i)} \varphi(m_{g(i)})$$

$$= \sum_{P \subset (N \cup \{0\} \backslash I)} (-1)^{\#P} \sum_{g \in \mathcal{F}(N \cup \{0\} \backslash I, \rightarrow P)} \prod_{i \in N \cup \{0\} \backslash I} \tilde{\ell}_{ig(i)} \varphi(m_{g(i)})$$

NOTATION **4.4.15**:
Let A be a finite set, and let g be a function of $\mathcal{F}(A,A)$. Then, Transit(g) denotes the set of subsets P of A which satisfy $g \in \mathcal{F}(A, \rightarrow P)$. Recall (Definition 2.4.6) that this means that for all a in A, there is a nonnegative integer k such that $g^k(a)$ is in P.

With this additional notation:

$$\sum_{g \in \mathfrak{F}(N\cup\{0\}\setminus I,\, \to I)} \prod_{i\in N\cup\{0\}\setminus I} \ell_{ig(i)}\varphi(m_{g(i)})$$

$$= \sum_{g \in \mathfrak{F}(N\cup\{0\}\setminus I,\, N\cup\{0\}\setminus I)} \left(\sum_{P \in \mathrm{Transit}(g)} (-1)^{\#P}\right) \prod_{i\in N\cup\{0\}\setminus I} \tilde{\ell}_{ig(i)}\varphi(m_{g(i)})$$

Let us assume for a while the following:

SUBLEMMA **4.4.16**:
Let A *be a finite set, and let* g *be an element of* $\mathfrak{F}$(A,A).

$$\sum_{P \in \mathrm{Transit}(g)} (-1)^{\#P} = \begin{cases} (-1)^{\#A}\mathrm{signature}(g) & \text{if } g \text{ is bijective,} \\ 0 & \text{otherwise.} \end{cases}$$

❑

Then:

$$\sum_{g \in \mathfrak{F}(N\cup\{0\}\setminus I,\, \to I)} \prod_{i\in N\cup\{0\}\setminus I} \ell_{ig(i)}\varphi(m_{g(i)})$$

$$= (-1)^{n+1+\#I} \prod_{i\in N\cup\{0\}\setminus I} \varphi(m_i) \sum_{g \in \sigma(N\cup\{0\}\setminus I)} \left(\mathrm{signature}(g) \prod_{i\in N\cup\{0\}\setminus I} \tilde{\ell}_{ig(i)}\right)$$

where $\sigma(N\cup\{0\}\setminus I)$ denotes the set of bijections of $N\cup\{0\}\setminus I$.
Lemma 4.4.9 is now clear if 0 belongs to I. Otherwise:

$$\sum_{g \in \sigma(N\cup\{0\}\setminus I)} \mathrm{signature}(g) \prod_{i\in N\cup\{0\}\setminus I} \tilde{\ell}_{ig(i)}$$

$$= \det(E(\mathbb{L}_{N\cup\{0\}\setminus I})) - (\ell_{00} - \tilde{\ell}_{00})\det(E(\mathbb{L}_{N\setminus I}))$$

where:

$$\ell_{00} - \tilde{\ell}_{00} = \frac{Lk_S(\mu_0,K_0)\varphi(m_0) + q_0\varphi(\mathfrak{L}_0)}{q_0\varphi(m_0)} = \frac{\varphi(\mu_0)}{q_0\varphi(m_0)} = \det(E(\mathbb{H}))$$

So, Lemma 4.4.9 is clear in both cases, assuming Sublemma 4.4.16.

❑

PROOF OF SUBLEMMA 4.4.16

Let g be a function of $\mathfrak{F}(A,A)$ where A is a finite set; g restricts to a bijection on $\bigcap_{k\in\mathbb{N}} g^k(A)$, this bijection decomposes into k cycles with disjoint supports A_1, A_2, ..., A_k. Let B denote the complement of $\bigcap_{k\in\mathbb{N}} g^k(A)$ in A.

$\{B, A_1, A_2, \ldots, A_k\}$ is a partition of A.

A subset P of A belongs to Transit(g) if and only if P intersects each of the A_i. Let $\mathfrak{P}(A)$ denote the set of subsets of A. So, Transit(g) is the subset of $\mathfrak{P}(B)\times\mathfrak{P}(A_1)\times \ldots \times\mathfrak{P}(A_k) = \mathfrak{P}(A)$ containing the $(P_0, P_1, \ldots, P_k)$ such that $p_i = \#P_i$ is positive if $i \neq 0$. Therefore,

$$\sum_{P\in \text{Transit}(g)} (-1)^{\#P} = \sum_{p_0=0}^{\#B} \binom{\#B}{p_0}(-1)^{p_0} \prod_{i=1}^{k} \sum_{p_i=1}^{\#A_i} \binom{\#A_i}{p_i}(-1)^{p_i}$$

where

$$\sum_{p_0=0}^{\#B} \binom{\#B}{p_0}(-1)^{p_0} = (1-1)^{\#B} = \begin{cases} 0 & \text{if B is not empty} \\ 1 & \text{otherwise} \end{cases}$$

and

$$\sum_{p_i=1}^{\#A_i} \binom{\#A_i}{p_i}(-1)^{p_i} = -1$$

Since B is empty if and only if g is bijective, and since the signature of a permutation of A which decomposes into k cycles with disjoint supports is $(-1)^{\#A+k}$, Sublemma 4.4.16 is proved.

❑

So are Lemma 4.4.9, and (hence) Lemma 4.2.3.

❑❑❑

§4.5 Proof of Lemma 4.2.5

Let $H = \{H_1, H_2, \ldots, H_b\}$ be a fixed b-component link in a rational homology sphere M (up to isotopy in M) such that:

For $i,j \in B = \{1, \ldots, b\}$, if $i\neq j$, then $Lk_M(H_i, H_j) \neq 0$.

Let $q_1, q_2, \ldots, q_b$ be b fixed positive integers.

Say that a b-tuple $R = (r_1, r_2, \ldots, r_b) \in \mathbb{Q}^b$ is *permissible* if, for $i \in B$, the ordered pair (r_i,q_i) specifies a primitive satellite $\mu_{r_i}(H_i)$ of H_i in M (see 1.3.1), that is if the number $(r_i - q_i Lk_M(H_i,H_i))$ (well-defined mod q_i) is an integer prime to q_i.
If R is a permissible b-tuple, $\mathbb{H}(R)$ denotes the surgery presentation in M:

$$\mathbb{H}(R) = (H_i\,;(r_i,q_i))_{i \in B} = (H_i\,;\mu_{r_i}(H_i))_{i \in B}$$

We want to show:

4.5.1 *For any permissible* $R \in \mathbb{Q}^b$, $F(M,\mathbb{H}(R))$ *is true.*

To do this, we first write M as $\chi(\mathbb{L}_N)$ where N is a finite set disjoint from B and $\mathbb{L}_N$ is a surgery presentation in S^3. Then we isotope H in M so that H is disjoint from the cores of the surgery performed on $\mathbb{L}_N$.
We denote by L_B the link H viewed in S^3, and by $\mathbb{L}_B(R)$ the surgery presentation $\mathbb{H}(R)$ viewed in S^3. (Here we think of a surgery presentation as a link equipped with characteristic curves.)
The links L_N and L_B are fixed from now on.

The formulae $F(S^3, \mathbb{L}_N)$ and $F(S^3, \mathbb{L}_{B\cup N}(R))$ are true. So, according to Lemma 4.2.2, in order to show that $F(M,\mathbb{H}(R))$ is true, it suffices to prove that $(*\mathbb{L}_{B\cup N}(R),N)$ is satisfied, that is, to prove:

$$P(R) = P(r_1, r_2, \ldots, r_b) = 0$$

where

$$\begin{aligned} P(R) = {} & \Big(\prod_{i\in B\cup N} q_i\Big) \sum_{\{I\,/\,I\neq\emptyset\,,\,I\subset B\cup N\}} \det(E(\mathbb{L}_{B\cup N\setminus I}(R)))\,\zeta_{B\cup N}(L_I) \\ & - \det(E(\mathbb{H}_B(R)))\Big(\prod_{i\in N\cup B} q_i\Big) \sum_{\{I\,/\,I\neq\emptyset\,,\,I\subset N\}} \det(E(\mathbb{L}_{N\setminus I}))\,\zeta_N(L_I) \\ & - \operatorname{sign}(\mathbb{L}_N)\Big(\prod_{i\in B} q_i\Big) \sum_{\{I\,/\,I\neq\emptyset\,,\,I\subset B\}} \det(E(\mathbb{H}_{B\setminus I}(R)))\,\zeta_B(H_I) \\ & - \frac{1}{24}\Big(\prod_{i\in B\cup N} q_i\Big)\sum_{i\in N} \frac{\det(E(\mathbb{L}_{B\cup N\setminus\{i\}}(R))) - \det(E(\mathbb{H}_B(R)))\det(E(\mathbb{L}_{N\setminus\{i\}}))}{q_i^2} \\ & - \frac{\det(E(\mathbb{L}_{B\cup N}(R)))}{24}\Big(\prod_{i\in B\cup N} q_i\Big)\sum_{i\in B}\left(\frac{Lk_{S^3}(\mu_{r_i}(H_i),H_i) - Lk_M(\mu_{r_i}(H_i),H_i)}{q_i}\right) \end{aligned}$$

Observe that P is a polynomial in the indeterminates $r_1, r_2, \ldots, r_b$.

(Recall that all the data except the r_i ($= Lk_M(H_i,\mu_{r_i}(H_i))$), for $i \in B$, are fixed, and note that $[Lk_M(H_i,\mu_{r_i}(H_i)) - Lk_{S^3}(H_i,\mu_{r_i}(H_i))]$ is independent of R.)
To prove Lemma 4.2.5, it suffices to show that our fixed polynomial P is zero in $\mathbb{Q}[r_1, r_2, \ldots, r_b]$. To do this, we construct a nonzero polynomial Q in the integral domain $\mathbb{Q}[r_1, r_2, \ldots, r_b]$ such that PQ is zero.

STEP 1
We construct a nonzero polynomial Q such that:
For any permissible $R \in \mathbb{Q}^b$, if $Q(R) \neq 0$, then:
For any $i \in B$, Property $\mathcal{P}(i)$ is satisfied, and,
for any $(i,j) \in B^2$ such that $j \leq i < b$, Property $\mathcal{P}(i,j)$ is satisfied with:
Property $\mathcal{P}(i)$: $M_i(R) = \chi_M(\mathbb{H}_{\{1, 2, \ldots, i\}}(R))$ is a rational homology sphere.
Property $\mathcal{P}(i,j)$: in $M_i(R)$, the linking number of H_{i+1} with the core $C_j(R)$ of the surgery performed on H_j is nonzero.

To do this, we associate with any property $\mathcal{P}$ of

$$\{\mathcal{P}(i) \ / \ i \in B\} \cup \{\mathcal{P}(i,j) \ / \ (i,j) \in B^2, j \leq i < b\}$$

a nonzero polynomial $Q(\mathcal{P}) \in \mathbb{Q}[r_1, r_2, \ldots, r_b]$, described below, such that:

$$Q(\mathcal{P})(R) \neq 0 \Rightarrow \mathcal{P} \text{ is satisfied.}$$

Then, we define Q as the product of these nonzero polynomials.

The manifold $\chi_M(\mathbb{H}_{\{1, 2, \ldots, i\}}(R))$ is a rational homology sphere if and only if

$$\det(E(\mathbb{H}_{\{1, 2, \ldots, i\}}(R))) \neq 0.$$

Set $Q(\mathcal{P}(i))(R) = \det(E(\mathbb{H}_{\{1, 2, \ldots, i\}}(R)))$.
As a polynomial in the indeterminates r_j, $j=1, \ldots, i$, $Q(\mathcal{P}(i))$ is nonzero because its term of highest total degree is:

$$\prod_{k \in \{1, 2, \ldots, i\}} \frac{r_k}{q_k}$$

In $H_1(M_i(R) \setminus H_{i+1};\mathbb{Q})$, if $j \leq i$, the core $C_j(R)$ and the meridian m_j of H_j in M are related by:

$$m_j = - q_j C_j(R)$$

So, the linking number in $M_i(R)$ of $C_j(R)$ ($j \leq i$) and H_{i+1} is zero if and only if m_j is a linear combination of the characteristic curves μ_k of the H_k ($k \leq i$) in $H_1(M \setminus \bigcup_{k=1}^{i+1} H_k \,;\, \mathbb{Q})$, that is, if and only if the minor of $E(\mathbb{H}_{\{1, 2, \ldots, i+1\}}(R))$ obtained by deleting the $(i+1)^{th}$ column and the jth row is zero.

This minor is a polynomial $Q(\mathbb{P}(i,j))$ of $\mathbb{Q}[r_1, r_2, \ldots, r_b]$. It is nonzero because its term of highest total degree is:

$$\pm \mathrm{Lk}_M(H_{i+1}, H_j) \prod_{k\in\{1,2,\ldots,i\}\setminus\{j\}} \frac{r_k}{q_k}$$

So, we have constructed the nonzero polynomial Q we were looking for.

STEP 2

We prove that PQ(R) is zero for any permissible $R \in \mathbb{Q}^b$. This is sufficient to conclude that PQ is zero and hence that Lemma 4.2.5 is true.

Let R be a permissible b-tuple such that $Q(R) \neq 0$.
Then, according to Lemmas 4.2.2 and 4.2.3, for any $i \geq 2$:

If $F(M, \mathbb{H}_{\{1,2,\ldots,i-1\}}(R))$ is true,

then $F(M, \mathbb{H}_{\{1,2,\ldots,i\}}(R))$ is equivalent to $F(M_{i-1}(R), \mathbb{H}_{\{i\}}(R) \subset M_{i-1}(R))$.

Since the surgery formula is true for the one-component surgery presentations in any rational homology sphere (4.2.4), $F(M,\mathbb{H}(R))$ is true by induction on i.
Now, (the converse of) Lemma 4.2.2 ensures that P(R) is zero.
So, PQ(R) is zero for any permissible R, and Lemma 4.2.5 is proved.

❑

REMARK **4.5.2**: Let λ_r be the invariant of rational homology spheres defined from the Walker invariant λ_W (see §4.6) by:

$$\lambda_r(M) = \frac{|H_1(M)|\lambda_W(M)}{2}$$

The Walker surgery formula can be shown to be equivalent to the formula T2 for λ_r and one-component surgery presentations of rational homology spheres. (This will be done in §4.7 for integral surgery presentations.) Then Lemmas 4.2.2 and 4.2.3 prove without requiring Chapter 3 that λ_r satisfies T2 for any surgery presentation $\mathbb{H}$ such that the minors of the linking matrix $E(\mathbb{H})$ described in Step 1 of the proof of Lemma 4.2.5 above are nonzero.

§4.6 The Walker surgery formula

Recall from [W] (Theorem 5.1):

THEOREM (Walker):

There exists a unique function λ_W:

{oriented rational homology spheres (up to orientation-preserving homeomorphisms)} $\to \mathbb{Q}$

such that:

W1. $\lambda_W(S^3) = 0$.

W2. For any one-component surgery presentation (K,μ) in a rational homology sphere M:

If $\chi_M(K,\mu)$ is a rational homology sphere:

$$\lambda_W(\chi_M(K,\mu)) = \lambda_W(M) + \frac{\langle m,\mu\rangle}{\langle m,\nu\rangle\ \langle\mu,\nu\rangle}\ \Delta_W''(M\backslash K)(1) + \tau_W(m,\mu;\nu)$$

where:

• $\langle\ ,\ \rangle$ is the intersection form of the boundary $\partial T(K)$ of the tubular neighborhood of K,

• m is the oriented meridian of K,

• μ is the characteristic curve of the surgery,

• ν is a generator of the kernel of the natural map from $H_1(\partial T(K);\mathbb{Z})$ to $H_1(M\backslash K;\mathbb{Z})$,

• $\Delta_W(M\backslash K) \in \mathbb{Q}[t^{1/2},t^{-1/2}]$ is the Alexander polynomial of $M\backslash K$, normalized so that:

The multiplication by t corresponds to the action of a generator of $H_1(M\backslash K)/\text{Torsion}$ on the infinite cyclic covering of $M\backslash K$, Δ_W is symmetric, and $\Delta_W(M\backslash K)(1) = 1$.

Thus:

$$\Delta_W(M\backslash K)(t)\ |\text{Torsion}(H_1(M\backslash K))| = \Delta(K\subset M)(t^{O_M(K)})$$

and,

$$\textbf{4.6.1} \qquad \Delta_W''(M\backslash K)(1) = \frac{O_M(K)^2\Delta''(K\subset M)(1)}{|\text{Torsion}(H_1(M\backslash K))|}$$

• if x and y are two elements of $H_1(\partial T(K);\mathbb{Z})$ such that

$$\langle x,y\rangle = 1 \text{ and } \nu = \delta y \text{ for some } \delta \in \mathbb{Z}:$$

$$\tau_W(m,\mu;\nu) = -\ \text{sign}\langle y,m\rangle s(\langle x,m\rangle,\langle y,m\rangle) + \text{sign}\langle y,\mu\rangle s(\langle x,\mu\rangle,\langle y,\mu\rangle) + \frac{(\delta^2-1)\ \langle m,\mu\rangle}{12\ \langle m,\nu\rangle\ \langle\mu,\nu\rangle}$$

❑

§4.7 Comparing T2 with the Walker surgery formula

Any rational homology sphere M can be obtained from S^3 after a finite number of integral surgeries on knots so that any intermediate 3-manifold is a rational homology sphere.

(PROOF: Let $\mathbb{L}$ be an n-component integral surgery presentation of M in S^3. Since $\mathbb{L}$ presents a rational homology sphere, the determinant of $E(\mathbb{L})$ is nonzero. So, $E(\mathbb{L})$ represents a non-degenerate bilinear symmetric form on

$$\left(\frac{\mathbb{Z}}{p\mathbb{Z}}\right)^n$$

for some odd prime integer p. Thus, $E(\mathbb{L})$ can be diagonalized over $\mathbb{Z}/p\mathbb{Z}$ using a change of basis of $\mathbb{Z}^n$. Since this change of basis may be realized by Kirby moves (see 6.3.5), it transforms $E(\mathbb{L})$ into the linking matrix $E(\mathbb{H})$ of an integral surgery presentation $\mathbb{H}$ of M. Now, if we perform the surgeries prescribed by $\mathbb{H}$ one by one, each intermediate manifold is a rational homology sphere. ❑)

So, to prove that the renormalized Walker invariant λ_r of rational homology spheres defined by

$$\lambda_r(M) = \frac{|H_1(M)|\lambda_W(M)}{2}$$

is equal to λ, it suffices to prove that λ_r satisfies Proposition T2 for one-component integral surgery presentations. This, according to Property W2, is equivalent to:

PROPOSITION **4.7.1**:
For any integral surgery presentation (K,μ) *in a rational homology sphere* M, *if* $\chi_M(K,\mu)$ *is a rational homology sphere, then:*

$$\frac{2\mathbb{F}_M(K,\mu)}{|H_1(\chi_M(K,\mu))|} = \frac{<m,\mu>}{<m,\nu>\,<\mu,\nu>}\,\Delta_W''(M\setminus K)(1) + \tau_W(m,\mu;\nu)$$

PROOF: Since (K,μ) is an integral surgery presentation,

$$\textbf{4.7.2} \qquad <m,\mu> = 1$$

By definition,

$$\textbf{4.7.3} \qquad |<m,\nu>| = O_M(K)$$

$Hom(H_1(M\setminus K;\mathbb{Z});\mathbb{Z})$ is isomorphic to $\mathbb{Z}$ and generated by the homomorphism which to any element of $H_1(M\setminus K;\mathbb{Z})$ associates its algebraic intersection number with an oriented surface of $M\setminus K$ with boundary ν. So,

$$\textbf{4.7.4} \qquad |H_1(\chi_M(K,\mu))| = |<\mu,\nu>|\;|Torsion(H_1(M\setminus K))|$$

$$\textbf{4.7.5} \qquad |H_1(M)| = |<m,\nu>|\;|Torsion(H_1(M\setminus K))|$$

Furthermore, by definition

$$\textbf{4.7.6} \qquad Lk_M(\mu,K) = \frac{<\mu,\nu>}{<m,\nu>}$$

So,

$$\textbf{4.7.7} \qquad sign(K,\mu) = sign\left(\frac{<\mu,\nu>}{<m,\nu>}\right)$$

Now, using Equations 4.7.2 to 4.7.7 and Definition 1.4.1 of $\zeta(K)$, Expression 1.7.3 of $\mathbb{F}_M(K,\mu)$ becomes:

$$\mathbb{F}_M(K,\mu) =$$

$$\mathrm{sign}\left(\frac{<\mu,\nu>}{<m,\nu>}\right)\left(\frac{O_M(K)}{2}\Delta''(K)(1) - |H_1(\chi_M(K,\mu))|\frac{|<m,\nu>|}{24|<\mu,\nu>|}\left(\frac{1}{<m,\nu>^2} + 1\right)\right)$$

$$+ |H_1(\chi_M(K,\mu))|\left(\frac{\mathrm{sign}\left(\frac{<\mu,\nu>}{<m,\nu>}\right)}{8} - \frac{<\mu,\nu>}{24 <m,\nu>}\right)$$

By 4.6.1, 4.7.2, 4.7.3 and 4.7.4, this may be rewritten as:

$$\frac{2\mathbb{F}_M(K,\mu)}{|H_1(\chi_M(K,\mu))|} = \frac{<m,\mu>}{<m,\nu> <\mu,\nu>}\Delta_W''(M\setminus K)(1)$$

$$- \frac{1}{12 <m,\nu> <\mu,\nu>} - \frac{<\mu,\nu>}{12 <m,\nu>} - \frac{<m,\nu>}{12 <\mu,\nu>} + \frac{\mathrm{sign}(<m,\nu> <\mu,\nu>)}{4}$$

The proof of the proposition is now reduced to the proof of 4.7.8.

4.7.8

$$\tau_W(m,\mu;\nu) =$$

$$- \frac{1}{12 <m,\nu> <\mu,\nu>} - \frac{<\mu,\nu>}{12 <m,\nu>} - \frac{<m,\nu>}{12 <\mu,\nu>} + \frac{\mathrm{sign}(<m,\nu> <\mu,\nu>)}{4}$$

PROOF OF 4.7.8:

Let x, y be such that <x,y> =1,

Writing

$$m = <m,y>x + <x,m>y \qquad \text{and} \qquad \mu = <\mu,y>x + <x,\mu>y,$$

shows

$$<m,\mu> = <m,y> <x,\mu> - <x,m> <\mu,y> = 1$$

Since the Dedekind sums satisfy:

- $s(q, -p) = s(q,p) = - s(-q,p)$,
- if q^{-1} is an inverse of q modulo p, then $s(q^{-1},p) = s(q,p)$,
- and the Dedekind reciprocity law (see 3.3.3),

we have:

$$- \mathrm{sign}<y,m>s(<x,m>,<y,m>) + \mathrm{sign}<y,\mu>s(<x,\mu>,<y,\mu>)$$

$$= - \mathrm{sign}<m,y>s(<\mu,y>,<m,y>) - \mathrm{sign}<\mu,y>s(<m,y>,<\mu,y>)$$

$$= - \frac{1}{12 <m,y> <\mu,y>} - \frac{<\mu,y>}{12 <m,y>} - \frac{<m,y>}{12 <\mu,y>} + \frac{\mathrm{sign}(<m,y> <\mu,y>)}{4}$$

If furthermore, $\nu = \delta y$, then

$$\tau_W(m,\mu;\nu) = - \mathrm{sign}<y,m>s(<x,m>,<y,m>) + \mathrm{sign}<y,\mu>s(<x,\mu>,<y,\mu>)$$

$$+ \frac{\delta^2-1}{12 <m,\nu> <\mu,\nu>}$$

$$\tau_W(m,\mu;\nu) =$$
$$-\frac{1}{12 <m,\nu> <\mu,\nu>} - \frac{<\mu,\nu>}{12 <m,\nu>} - \frac{<m,\nu>}{12 <\mu,\nu>} + \frac{\mathrm{sign}\left(<m,\nu> <\mu,\nu>\right)}{4}$$

This proves 4.7.8 and completes the proof of 4.7.1.

❑

We have thus proved

PROPERTY **T5.0**:

For any rational homology sphere M

$$\lambda(M) = \frac{|H_1(M)|\lambda_W(M)}{2}$$

❑

REMARK **4.7.9**: We may now delete the hypothesis "integral" in the statement of 4.7.1. This allows us to describe the Walker function τ_W by the formula:

$$\tau_W(m,\mu;\nu) = s(<\mu,\ell> , <m,\mu>)\ \mathrm{sign}(<m,\mu>) + \frac{\mathrm{sign}\left(<m,\mu> <m,\nu> <\mu,\nu>\right)}{4}$$
$$- \frac{<m,\mu>}{12 <m,\nu> <\mu,\nu>} - \frac{<\mu,\nu>}{12 <m,\mu> <m,\nu>} - \frac{<m,\nu>}{12 <\mu,\nu> <m,\mu>}$$

where ℓ denotes an element of $H_1(\partial T(K);\mathbf{Z})$ such that $<m,\ell> = 1$ (with the notation of §4.6).

Chapter 5

The invariant λ for 3-manifolds with nonzero rank

§5.1 Introduction

In this section, we compute $\lambda(M)$ for any oriented closed 3-manifold M with positive rank. (The rank of a closed 3-manifold M is its first Betti number $\beta_1(M)$.) In order to do this, we first give M a surgery presentation as in:

LEMMA **5.1.1**:
Any oriented closed 3-manifold M *can be obtained by surgery from a rational homology sphere* R *according to the instructions of a presentation* $\mathbb{L}$ *such that: The linking matrix of* $\mathbb{L}$ *is null and the components of the underlying link* L *of* $\mathbb{L}$ *are null-homologous.*
The link L *has then* $\beta_1(M)$ *components, and*

$$|H_1(R;\mathbb{Z})| = |\mathrm{Torsion}(H_1(M;\mathbb{Z}))|$$

PROOF: Let $\beta = \beta_1(M)$. Let $C = \{C_1, ..., C_\beta\}$ be a link in M such that the components of C represent a basis of $H_1(M)/\mathrm{Torsion}$.
Let $S_1, ..., S_\beta$ be β oriented surfaces embedded in M such that S_i intersects C_i exactly once transversally and S_i does not intersect C_j if $i\neq j$. (This geometric realization of Poincaré duality is outlined in the proof of Consequence 5.2.3.)
The meridian μ_i of C_i bounds a surface Σ_i (the surface S_i minus a meridian disk of C_i) in $M\setminus C$.
The inclusion induces an isomorphism from $H_1(M\setminus C;\mathbb{Z})$ to $H_1(M;\mathbb{Z})$.
Let m_i be a parallel of C_i in the boundary of the tubular neighborhood $T(C_i)$ of C_i.
Performing surgery on M with respect to the surgery presentation $(C_i,m_i)_{i=1, ..., \beta}$ gives a 3-manifold R with

$$H_1(R;\mathbb{Z}) = \frac{H_1(M\backslash C;\mathbb{Z})}{\bigoplus_{i=1}^{\beta} \mathbb{Z}m_i}$$

So, $H_1(R)$ is a torsion group with the same torsion as $H_1(M)$, and R is thus a rational homology sphere.
Let K_i denote the core of the surgery performed on C_i in R, then the surgery presentation $\mathbb{L} = (K_i,\mu_i)_{i=1,\dots,\beta}$ in R satisfies all the properties required in the statement of the lemma. (The longitude μ_i of K_i bounds the Seifert surface Σ_i in $R\backslash(K_j)_{j=1,\dots,\beta}$.)

❑

Next, we apply the surgery formula T2 to write

5.1.2 $$\lambda(M) = \mathbb{F}_R(\mathbb{L}) = \begin{cases} \frac{1}{2}\Delta''(L)(1) - \frac{|H_1(R)|}{12} & \text{if } \beta_1(M) = 1 \\ \zeta(L) & \text{if } \beta_1(M) \geq 2 \end{cases}$$

Now, the case "$\beta_1(M) = 1$" follows easily from

LEMMA **5.1.3**:
If $\beta_1(M)$ *equals one, with the notation of Lemma* 5.1.1, *the Alexander polynomials* $\Delta(M)$ *and* $\Delta(L\subset R)$ *are the same.*

❑

PROOF OF LEMMA 5.1.3: Since the inclusion induces an isomorphism from $H_1(R\backslash L;\mathbb{Z})$ to $H_1(M;\mathbb{Z})$, the infinite cyclic covering $\tilde{X}(L)$ of the exterior of L is injected into the infinite cyclic covering $\tilde{M}$ of M, and this injection is compatible with the action of $\frac{H_1(M)}{\text{Torsion}} \cong \frac{H_1(R\backslash L)}{\text{Torsion}} \cong \mathbb{Z}$.

($\tilde{M} \setminus \tilde{X}(L)$) is an infinite solid cylinder and its meridian bounds, in $\tilde{X}(L)$, a lifting of a Seifert surface of the knot L in R. So, $H_1(\tilde{X}(L);\mathbb{Z})$ and $H_1(\tilde{M};\mathbb{Z})$ are isomorphic $\mathbb{Z}[t,t^{-1}]$-modules, and thus, according to Proposition 2.3.13,

$$\Delta(M) = \Delta(L\subset R)$$

❑

We have thus proved:

PROPERTY T5.1:
Let M *be an oriented closed 3-manifold with rank* $\beta_1(M) = 1$, *then*

$$\lambda(M) = \frac{1}{2}\Delta''(M)(1) - \frac{|\mathrm{Torsion}(H_1(M))|}{12}$$

❑

We are now left with the computation of $\lambda(M)$ for oriented closed 3-manifolds M with $\beta_1(M) \geq 2$.

DEFINITION **5.1.4**: *Homology unlink*
A link L in a manifold R will be called a *homology unlink* if any component of L is null-homologous and if any two components of L have linking number 0.

Let L be an oriented null-homologous link of $n\geq 2$ components in a rational homology sphere R, then according to 2.1.1, 2.3.13, 2.3.14 and 2.3.15,
5.1.5

$$\mathcal{D}(L)(u, u, \ldots, u) = |H_1(R)|\, z^{n-2}(a_0(L) + a_1(L)z^2 + \ \ldots + a_{d(L)}(L)z^{2d(L)})$$

with

$$z = \exp(\frac{u}{2}) - \exp(- \frac{u}{2})$$

A quick look at the Taylor series of $\mathcal{D}(L)(u, u, \ldots, u)$ at $u = 0$, as deduced from 5.1.5 and from 2.5.2 justifies the following lemma.

LEMMA **5.1.6**:
If L *is a homology unlink with at least 2 components in a rational homology sphere* R:

$$\frac{\zeta(L)}{|H_1(R)|} = a_1(L)$$

❑

CONSEQUENCE **5.1.7**:
Let M *be an oriented closed 3-manifold with rank* $\beta_1(M) \geq 2$, *then with the notation of Lemma* 5.1.1,

$$\frac{\lambda(M)}{|\mathrm{Torsion}(H_1(M))|} = a_1(L)$$

❑

Now, we are only left with the task of studying the a_1-coefficients of homology unlinks with at least 2 components and of relating them (in §5.3) to known invariants of the manifolds presented by such links equipped with zero surgery coefficients.

The study of the a_1-coefficients of homology unlinks has been performed by Hoste. His results will be recalled and reproved in §5.2 which is essentially an excerpt from [Hos].
The computation of λ for manifolds with rank greater than 1 will then be performed in §5.3.

§5.2 The coefficients a_1 of homology unlinks in rational homology spheres (after Hoste)

We first recall the following modification of an oriented surface which does not change its homology class.

DEFINITION **5.2.1**: *Addition of a tube to an oriented surface*
Let Σ be an oriented surface. Let α be an oriented arc intersecting Σ exactly at its ends $\alpha(0)$ and $\alpha(1)$ in the interior of Σ normally and with opposite signs. Consider a tubular neighborhood $N(\alpha)$ of α which intersects Σ exactly at open regular neighborhoods $N(\alpha(0))$ and $N(\alpha(1))$ of $\alpha(0)$ and $\alpha(1)$. We denote by $\Sigma(\alpha)$ the surface obtained from Σ as follows:

$$\Sigma(\alpha) = \Big(\Sigma \cup \partial \overline{N}(\alpha) \Big) \setminus \big(N(\alpha(0)) \cup N(\alpha(1)) \big)$$

and we call the surgery transforming Σ into $\Sigma(\alpha)$ an "*addition of a tube with core* α".

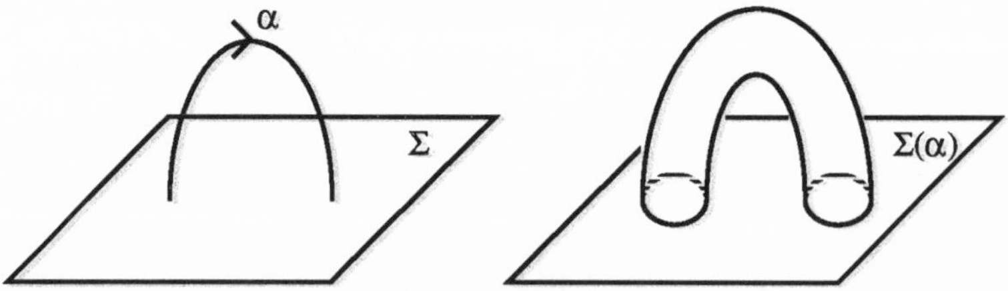

Figure 5.1

LEMMA **5.2.2**: *Connecting the intersection between two Seifert surfaces*
Let Σ_1 and Σ_2 be two connected compact oriented surfaces with non-empty boundaries embedded in general position in a closed 3-manifold R, *and such that they intersect only in their respective interiors. (Their intersection is thus a finite family of closed curves.)*
Then, it suffices to add to Σ_1 and Σ_2 a finite number of tubes to transform them into two surfaces with connected intersection, which still satisfy the initial assumptions.

PROOF: The proof is an induction on the number of connected components of $\Sigma_1 \cap \Sigma_2$.

If there are several curves in $\Sigma_1 \cap \Sigma_2$, any curve c_1 of $\Sigma_1 \cap \Sigma_2$ can be joined to another curve c_2 of $\Sigma_1 \cap \Sigma_2$ on the (connected) surface Σ_1 by a path α whose interior does not meet Σ_2.

Case 1: If α is on the same side of Σ_2 near both of its ends (that is, if the signs associated with the intersection of Σ_2 and α are opposite at the ends of α), a tube with core α may be added to Σ_2; and this operation changes $\Sigma_1 \cap \Sigma_2$ by transforming c_1 and c_2 into their connected sum along a band with core α in Σ_1. In particular, it reduces the number of components of the intersection.

Case 1: Adding a tube to Σ_2 to reduce $\Sigma_1 \cap \Sigma_2$

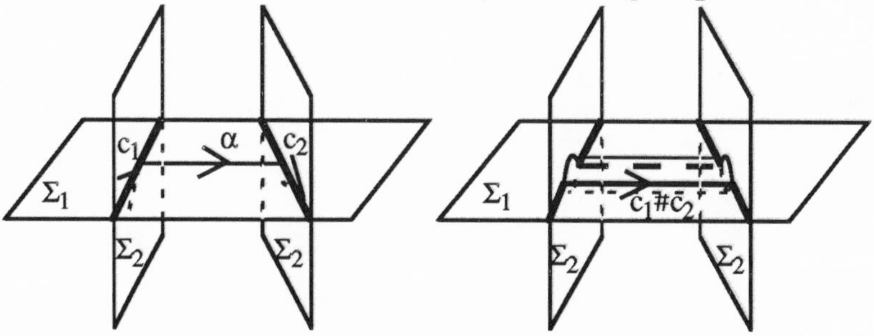

Figure 5.2

Case 2: The reduction of Case 1 cannot be performed.

Let c_1 be a component of $\Sigma_1 \cap \Sigma_2$ which can be joined to $\partial\Sigma_2$ by a path β in Σ_2 whose interior does not meet $\Sigma_1 \cap \Sigma_2$. Add to Σ_1 a tube with core γ, where γ is an arc in $R\backslash(\Sigma_1 \cup \Sigma_2)$ near β as in Figure 5.3. This tube does not meet Σ_2, so this operation does not change $\Sigma_1 \cap \Sigma_2$ and the reduction of Case 1 can now be performed.

❑

CONSEQUENCE **5.2.3**: *Preparing Seifert surfaces in order to compute the a_I-coefficients in Lemma* 5.2.5

Let $L = (K_1, ..., K_n)$ $(n \geq 2)$ *be a null-homologous oriented link in a rational homology sphere* R *such that* K_1 *is null-homologous in* $R\backslash(L\backslash K_1)$.

Then, there exist two (connected, oriented) Seifert surfaces $\Sigma_{>1}$ *and* Σ_1 *for* $L\backslash K_1$ *and* K_1 *such that* $\Sigma_{>1} \cap \Sigma_1$ *is an arc* γ *with ends* A *and* B *on* K_2.

Case 2 in the proof of Lemma 5.2.2:
Adding a tube to Σ_1 to satisfy the hypotheses of Case 1

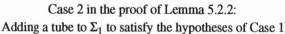

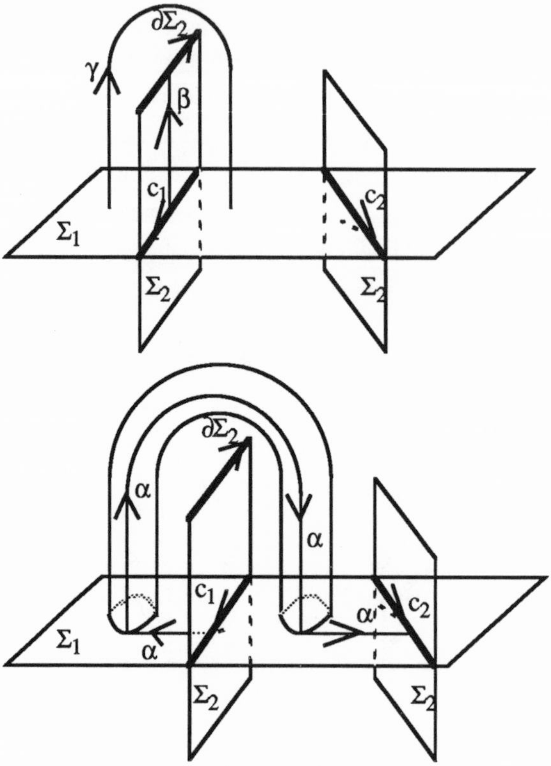

Figure 5.3

PROOF OF CONSEQUENCE 5.2.3: Since K_1 is null-homologous in $R\backslash(L\backslash K_1)$, it bounds a Seifert surface Σ_1 in $R\backslash(L\backslash K_1)$ and K_1 does not link algebraically the components of $L\backslash K_1$. Then a Seifert surface $\Sigma_{>1}$ for $(L\backslash K_1)$ in R does not intersect algebraically K_1 and can thus be made disjoint from K_1. (It suffices to remove successively consecutive intersection points which algebraically cancel by adding tubes with segments of K_1 as cores to $\Sigma_{>1}$.)

Now, Σ_1 and $\Sigma_{>1}$ intersect only in their interior and their intersection can be made connected by Lemma 5.2.2. Then Σ_1 can be isotoped so that the intersection circle c, as a curve in Σ_1, moves in $\Sigma_{>1}$, until it reaches K_2 and passes beyond it, so that the intersection of the surfaces becomes an arc γ. See Figure 5.4. (Note that the result is also true starting with disjoint Seifert surfaces Σ_1 and $\Sigma_{>1}$.)

❑

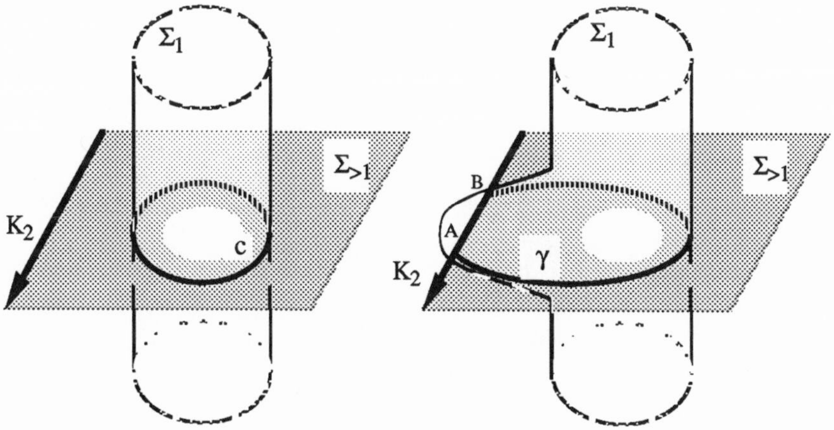

Figure 5.4

NOTATION **5.2.4**
The *linking matrix* Lk(L) of a (non-framed, oriented) link $L = (K_1, \ldots, K_n)$ in an oriented rational homology sphere is the matrix $Lk(L) = [\ell_{ij}]_{1 \leq i,j \leq n}$ where

$$\ell_{ij} = \begin{cases} Lk(K_i,K_j) & \text{if } i \neq j \\ -\sum_{k \in \{1, \ldots, n\} \setminus \{i\}} Lk(K_k,K_i) & \text{if } i=j \end{cases}$$

$Lk_{ii}(L)$ will denote the submatrix of Lk(L) obtained from Lk(L) by deleting its ith row and column.

LEMMA **5.2.5** (Hoste): *A way of computing* $a_1(L)$

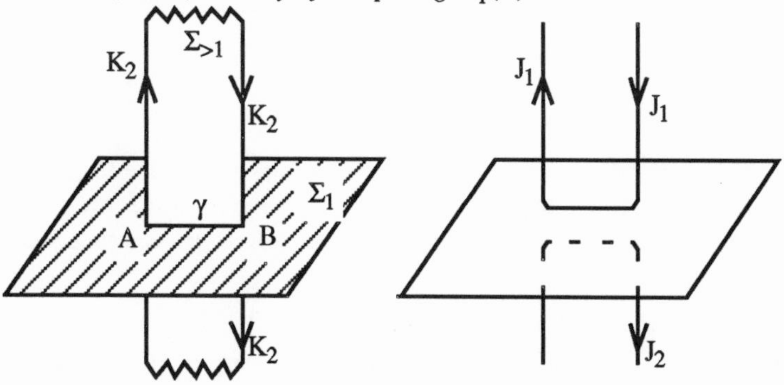

Figure 5.5

Let $L = (K_1, \ldots, K_n)$ $(n \geq 2)$ *be a null-homologous oriented link in an oriented rational homology sphere* R *such that* K_1 *is null-homologous in* $R \setminus (L \setminus K_1)$, *let* $\Sigma_{>1}$ *and* Σ_1 *be two (connected, oriented) Seifert surfaces for* $L \setminus K_1$ *and*

K_1 *such that* $\Sigma_{>1} \cap \Sigma_1$ *is an arc* γ *with ends on* K_2. *(Such surfaces exist by Consequence* 5.2.3.*)*
Let Σ' *be the surface obtained from* $\Sigma_{>1}$ *by cutting out a regular neighborhood of* γ. *Call* $L' = (J_1, J_2, K_3, ..., K_n)$ *its oriented boundary. (See Figure 5.5.) Then*

$$a_1(L) = - \det(Lk_{11}(L'))$$

PROOF: (contained in the proof of Lemma 1.4. in [Hos])
We may and do assume that γ does not separate $\Sigma_{>1}$.
Call e a loop of $\Sigma_{>1}$ intersecting γ once, then

$$H_1(\Sigma_{>1}) = \mathbb{Z}e \oplus H_1(\Sigma')$$

Construct a Seifert surface Σ for L from Σ_1 and $\Sigma_{>1}$ so that Σ is the union of Σ_1 and $\Sigma_{>1}$ outside a neighborhood of γ, and Σ is as in Figure 5.6 in the neighborhood of γ:

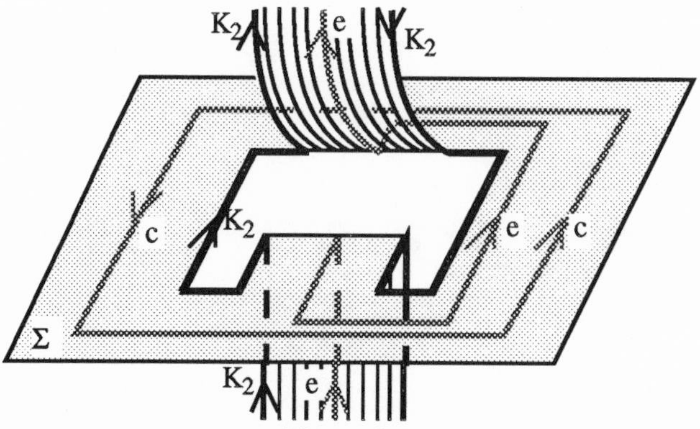

Figure 5.6

5.2.6 $\quad H_1(\Sigma) = \mathbb{Z}c \oplus \mathbb{Z}e \oplus H_1(\Sigma') \oplus H_1(\Sigma_1)$

Equip $H_1(\Sigma_1)$ with a symplectic basis (for the intersection form) and $H_1(\Sigma')$ with a basis formed with the classes of $J_2, K_3, ..., K_n$ followed by a symplectic system for the intersection form on Σ'.
Let $V(\Sigma)$ be the Seifert matrix associated with the basis induced by Decomposition 5.2.6 of $H_1(\Sigma)$ and the given bases of $H_1(\Sigma')$ and $H_1(\Sigma_1)$. According to 2.3.14,

$$\nabla_L(z = t^{1/2} - t^{-1/2}) = \det(t^{1/2}V(\Sigma) - t^{-1/2}\,{}^tV(\Sigma))$$

In this basis of $H_1(\Sigma)$, c (and c^+ as well) links only e and links it exactly once. So,

5.2.7

$$\nabla_L(z = t^{1/2} - t^{-1/2}) = - z^2 \left(\det(t^{1/2}V(\Sigma'\cup\Sigma_1) - t^{-1/2}\, {}^tV(\Sigma'\cup\Sigma_1))\right)$$

where $V(\Sigma'\cup\Sigma_1)$ is the matrix of the restriction of the Seifert form of Σ to $H_1(\Sigma') \oplus H_1(\Sigma_1)$.

$$t^{1/2}V(\Sigma'\cup\Sigma_1) - t^{-1/2}\, {}^tV(\Sigma'\cup\Sigma_1) = zV(\Sigma'\cup\Sigma_1) + t^{-1/2} (V(\Sigma'\cup\Sigma_1) - {}^tV(\Sigma'\cup\Sigma_1))$$

$(V(\Sigma'\cup\Sigma_1) - {}^tV(\Sigma'\cup\Sigma_1))$ is the matrix of the intersection form on $\Sigma'\cup\Sigma_1$. So its (n-1) first rows (corresponding to J_2, K_3, ..., K_n) as well as its (n-1) first columns are zero and the matrix obtained by deleting these rows and columns is symplectic.

In particular the (n-1) first columns of $\left(t^{1/2}V(\Sigma'\cup\Sigma_1) - t^{-1/2}\, {}^tV(\Sigma'\cup\Sigma_1)\right)$ can be divided by z. This division yields a matrix A(t) which has the (n-1) first columns of $V(\Sigma'\cup\Sigma_1)$ followed by the columns with rank greater than (n-1) of

$$\left(zV(\Sigma'\cup\Sigma_1) + t^{-1/2} (V(\Sigma'\cup\Sigma_1) - {}^tV(\Sigma'\cup\Sigma_1))\right)$$

and, according to 5.2.7,

$$\nabla_L(z) = - z^{n+1} \det(A(t))$$

So,

$$a_1(L) = - \det(A(1))$$

where

$$A(1) = \begin{pmatrix} Lk_{11}(L') & 0 & 0 & \ldots & 0 \\ ? & J & 0 & \ldots & 0 \\ ? & 0 & J & \ldots & 0 \\ ? & 0 & 0 & \ldots & 0 \\ ? & 0 & 0 & \ldots & J \end{pmatrix}$$

with $J = \begin{pmatrix} 0 & 1 \\ -1 & 0 \end{pmatrix}$ and the "0" represent null matrices of any order.

Thus

$$\det(A(1)) = \det(Lk_{11}(L'))$$

This proves Lemma 5.2.5.

❑

REMARK **5.2.8**: The arguments of the proof of 5.2.5 also prove:
For any null-homologous link L *in a rational homology sphere* R:

$$a_0(L) = \det(Lk_{ii}(L)), \textit{ for any } i = 1, \ldots, n.$$

This expression of $a_0(L)$ can also be deduced from 2.3.13 and from the Taylor series of $\mathcal{D}(L)(u, \ldots, u)$ given by 2.5.2.

LEMMA **5.2.9** (Hoste)
Let L *be an n-component homology unlink in a rational homology sphere.*
If $n \geq 4$,

$$a_1(L) = 0$$

PROOF: Derive a link L' from L, as in Lemma 5.2.5. So,

$$a_1(L) = -\det(Lk_{11}(L'))$$

Since, for $i \geq 3$,

$$Lk(J_1,K_i) + Lk(J_2,K_i) = Lk(K_2,K_i) = 0$$

$Lk_{11}(L')$ is an $(n-1)\times(n-1)$-matrix with rank at most 2. Thus, $Lk_{11}(L')$ has determinant zero and Lemma 5.2.9 is proved.

❑

§5.3 Computing λ for manifolds with rank at least 2

Lemmas 5.1.1, 5.1.7 and 5.2.9 prove

PROPOSITION **T5.≥4**

If M *is a closed oriented 3-manifold with rank* $\beta_1(M) \geq 4$, *then:*

$$\lambda(M) = 0$$

❑

Now, it only remains to prove the two propositions:

PROPOSITION **T5.2**

Let M *be an oriented closed 3-manifold with rank* $\beta_1(M) = 2$.
Let S_1 *and* S_2 *be two embedded oriented surfaces, in general position in* M, *such that the homology classes of* S_1 *and* S_2 *generate* $H_2(M;\mathbf{Z})$; *let* c *be their (possibly disconnected) oriented intersection, and let* c' *be the parallel to* c *inducing the trivialization of the tubular neighborhood of* c *given by the surfaces, then*

$$\frac{\lambda(M)}{|\text{Torsion}(H_1(M))|} = -Lk_M(c,c')$$

PROPOSITION **T5.3**

Let M *be an oriented closed 3-manifold with rank* $\beta_1(M) = 3$.
Let $\{a_1, a_2, a_3\}$ *be a basis of* $H^1(M;\mathbf{Z})$, *let* $\cup$ *denote the cup product and let* $[M] \in H_3(M;\mathbf{Z})$ *be the orientation class of* M, *then*

$$\frac{\lambda(M)}{|\text{Torsion}(H_1(M))|} = \left((a_1 \cup a_2 \cup a_3)([M])\right)^2$$

Preliminaries shared by the proofs of Propositions T5.2 and T5.3

Let M be a closed 3-manifold with rank n = 2 or 3, presented by a surgery presentation $\mathbb{L}$ in a rational homology sphere R with a null linking matrix and

with the homology unlink $L = (K_1, \ldots, K_n)$ as underlying link as in Lemma 5.1.1.
Let $\Sigma_{>1}$ and Σ_1 be two (connected, oriented) Seifert surfaces for $L \backslash K_1$ and K_1 such that $\Sigma_{>1}$ and Σ_1 intersect transversally along a connected curve c lying in the interior of both $\Sigma_{>1}$ and Σ_1. (See Lemma 5.2.2.)

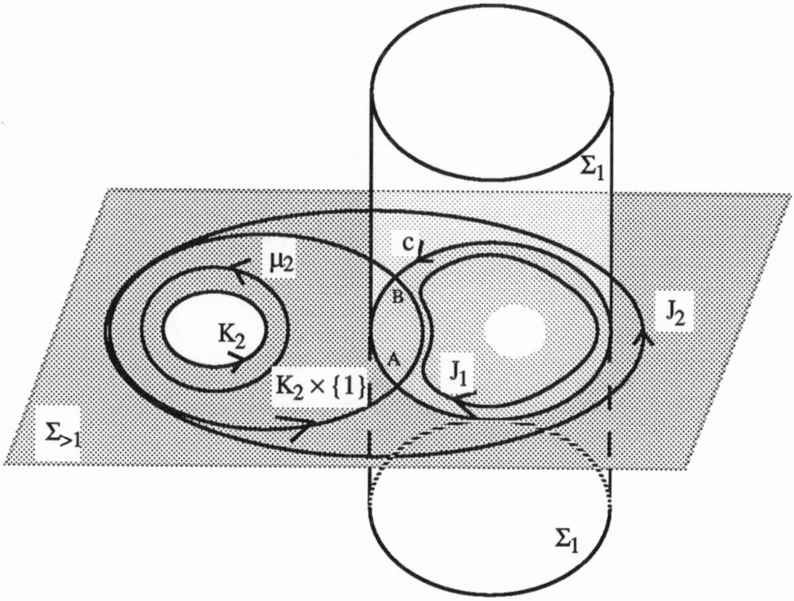

Figure 5.7: In R, a neighborhood of c, of K_2, and of a path joining them in $\Sigma_{>1}$ $(c = \Sigma_1 \cap \Sigma_{>1})$

Let $L \times D^2$ be a tubular neighborhood of L (D^2 is parametrized by the unit disk of $\mathbb{C}$) such that

- $\Sigma_{>1} \cap (L \times D^2) = (L \backslash K_1) \times [0,1]$
- $\Sigma_1 \cap ((L \backslash K_2) \times D^2) = K_1 \times [0,1]$
- $\Sigma_1 \cap (K_2 \times D^2)$ is a disk in $K_2 \times D^{2+}$ where $D^{2+} = \{z \in D^2, \mathrm{Re}(z) > \frac{1}{2}\}$, and
- Σ_1 intersects $K_2 \times [0,1]$ along a simple arc $c \backslash \mathring{\gamma}$.

The link $(K_1, K_2 \times \{1\}, K_3, \ldots, K_n)$ (isotopic to L) and the Seifert surfaces Σ_1 and $\Sigma_{>1} \backslash (K_2 \times [0,1[)$ satisfy the hypotheses of Lemma 5.2.5. So we can use them to define J_1, J_2 and L' as in Lemma 5.2.5 (see Figure 5.7) and to get

$$a_1(L) = -\det(Lk_{11}(L')) = \begin{cases} Lk_R(J_1, J_2) & \text{if } \beta_1(M) = 2 \\ Lk_R(J_2, K_3)^2 & \text{if } \beta_1(M) = 3 \end{cases}$$

Now, let us compute these linking numbers in M.

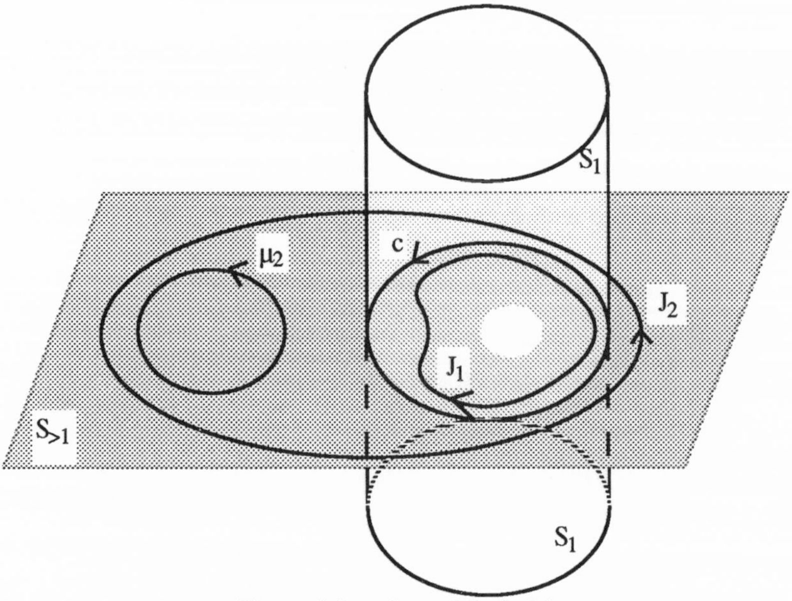

Figure 5.8: c, J_1, and J_2, in M

The surgery on $\mathbb{L}$ which transforms R into M is performed by removing $L\times\frac{1}{2}\mathring{D}^2$ from R and by gluing back n solid tori with meridians $\mu_i=K_i\times\frac{1}{2}$, i=1, ..., n.

5.3.1 Replacing each $K_i\times[0,\frac{1}{2}[$ by a meridian disk of a new solid torus with boundary $K_i\times\frac{1}{2}$ in the surfaces $\Sigma_{>1}$ and Σ_1 transforms them into closed surfaces $S_{>1}$ and S_1 in M which intersect along c.
J_1 and J_2 can now be seen in M as two oppositely oriented curves parallel to c (= $S_1 \cap S_{>1}$) and located on either side of c in $S_{>1}$.

PROOF OF PROPOSITION T5.2 (assuming that the mentioned linking number Lk(c,c') is a well-defined invariant of oriented closed 3-manifolds with rank 2)
Let S_1 and $S_2=S_{>1}$ be as in 5.3.1, by construction they form a basis of $H_2(M;\mathbb{Z})$. The pair ($c=S_1\cap S_2$,c') is isotopic (in M) to (J_1, J_2) up to orientation, and the orientation of one of the curves J_1 and J_2 is opposite to the orientation of c.
Since J_1 links neither K_1 nor K_2, J_1 bounds a surface in R\L, which we may see in M. So, since J_2 does not meet L, $Lk_M(J_1, J_2)$ and $Lk_R(J_1, J_2)$ are equal. This (together with 5.1.7) proves the proposition provided Lk(c,c') does not depend on the choice of S_1 and S_2. ❑

PROOF OF PROPOSITION T5.3

$Lk_R(J_2, K_3)$ is the intersection number of J_2 and a Seifert surface Σ_3 of K_3 in $R\backslash(L\backslash K_3)$, so $Lk_R(J_2, K_3)$ is also the intersection number of J_2 and S_3 in M. (We use the notation of the proof of 5.1.1, the surface S_3 is obtained from Σ_3 by replacing $K_3\times[0,\frac{1}{2}[$ by a meridian disk of C_3 in M.)

Let D denote the Poincaré duality isomorphism

$$D: H^*(M;\mathbb{Z}) \rightarrow H_{3-*}(M;\mathbb{Z})$$

and let the surfaces and the curves be (again) identified with their homology classes.

$$Lk_R(J_2,K_3) = \pm (D^{-1}(J_2) \cup D^{-1}(S_3))([M])$$

where

$$D^{-1}(J_2) = \pm D^{-1}(S_1) \cup D^{-1}(S_{>1})$$

and

$$S_{>1} = S_2 + S_3$$

So,

$$Lk_R(J_2,K_3) = \pm (D^{-1}(S_1) \cup D^{-1}(S_2 + S_3) \cup D^{-1}(S_3))([M])$$
$$= \pm (D^{-1}(S_1) \cup D^{-1}(S_2) \cup D^{-1}(S_3))([M])$$

and, since $\{S_1, S_2, S_3\}$ is dual to the generating system $\{C_1, C_2, C_3\}$ of

$$\frac{H_1(M;\mathbb{Z})}{\text{Torsion}}$$

for the intersection form, $(D^{-1}(S_1), D^{-1}(S_2), D^{-1}(S_3))$ is a basis of $H^1(M;\mathbb{Z})$.

❑

COMPLETING THE PROOF OF PROPOSITION T5.2

Let S_1 and S_2 be two closed, embedded surfaces (in general position) in M whose homology classes $[S_1]$ and $[S_2]$ generate $H_2(M;\mathbb{Q})$. Let c be the (possibly disconnected) oriented intersection of S_1 and S_2 and let c' be the parallel to c which induces the same framing of c as the surfaces do. By construction c and c' are rationally null-homologous in M, hence $Lk_M(c,c')$ is well-defined. We are about to prove that $Lk_M(c,c')$ is a quadratic function of the exterior product $[S_1]\wedge[S_2] \in H_2(M;\mathbb{Z})\wedge H_2(M;\mathbb{Z})$: This will imply in particular that the $Lk_M(c,c')$ in the statement of Proposition T5.2 does not depend on the choices of S_1 and S_2 and this will complete the proof of Proposition T5.2.

Firstly, since $Lk_M(c,c')$ is the linking number of c with its pushed-off c' in the positive normal direction of S_1, $Lk_M(c,c')$ only depends on S_1 and on the homology class of c in $H_1(S_1;\mathbb{Z})$. Secondly, the homology class of c in $H_1(S_1;\mathbb{Z})$ is well-determined by the homology class of S_2 in $H_2(M;\mathbb{Z})$ because, for a curve of S_1, its algebraic intersection with c in S_1 is (up to a well-determined sign) its algebraic intersection with S_2 in M. Thus $Lk_M(c,c')$ is

a well-determined function of S_1 and $[S_2]$. Similarly, $Lk_M(c,c')$ only depends on $[S_1]$. We denote it by $q_M([S_1],[S_2])$.
Now, it suffices to verify that for any two systems $([S_1],[S_2])$ and $([S'_1],[S'_2])$ of linearly independent elements of $H_2(M;\mathbb{Z})$ (with an obvious abuse of notation):

$$q_M([S'_1],[S'_2]) = \left(\frac{[S'_1]\wedge[S'_2]}{[S_1]\wedge[S_2]}\right)^2 q_M([S_1],[S_2])$$

It is sufficient to verify it if the two systems are obtained one from the other by an elementary operation for which it is easily verified.

❑

Chapter 6

Applications and variants of the surgery formula

Subsections 6.1 to 6.4 are independent. §6.1 and §6.3 present applications of the surgery formula while §6.2 and §6.4 prove the equivalence between Definition 1.4.8 of the surgery function $\mathbb{F}$ and the definitions of $\mathbb{F}$ given in §1.7.

§6.1 Computing λ for all oriented Seifert fibered spaces using the formula

6.1.A Introduction and statement of the result

§6.1 is devoted to computing λ for all oriented Seifert fibered spaces using the formula. These spaces are described in [Seif] and [Mo] and by the surgery presentations (taken from [Mo] Fig.12 p.146) of Figures 6.1 and 6.2 below.
The Casson-Walker invariant of Seifert fibered rational homology spheres has already been computed in [L1] (it fortunately gave the same result); and the definition of λ given by T5.>1 allows for a more geometric computation of λ for Seifert fibered spaces with rank greater than 1 (6.1.C shows that it also gives consistent results). So, this subsection is merely an example of concrete utilization of the formula.

REMARKS:
• [L1] generalized a formula obtained independently by Neumann and Wahl, and Fukuhara, Matsumoto and Sakamoto, to describe the Casson invariant of Seifert fibered integral homology spheres (see [N-W] and [F-M-S]).
• In this particular case, it is simpler to use the surgery formula T1 to compute the Walker invariant than to use the inductive process based on the Walker formula as in [L1]; but in general it is more convenient to use the algorithm described in [L2] to compute λ for rational homology spheres presented by surgery diagrams.

The oriented Seifert fibered space with an oriented genus g surface as its base and with n exceptional fibers (Oog |b; $(\alpha_i,\beta_i)_{i=1,\ldots,n}$) ($\alpha_i, \beta_i \in \mathbb{N}$, $0<\beta_i<\alpha_i$, g.c.d.$(\alpha_i,\beta_i) = 1$) is presented by the surgery diagram shown in Figure 6.1 below.

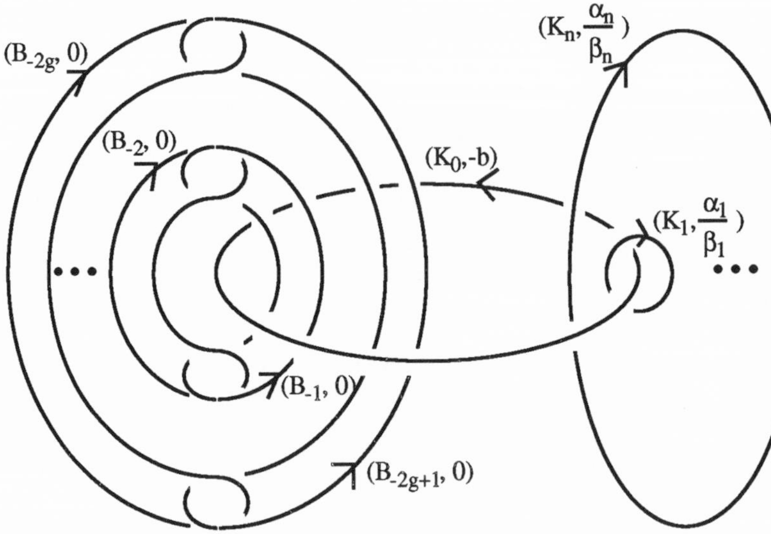

Figure 6.1: Surgery presentation of (Oog |b; $(\alpha_i,\beta_i)_{i=1,\ldots,n}$)

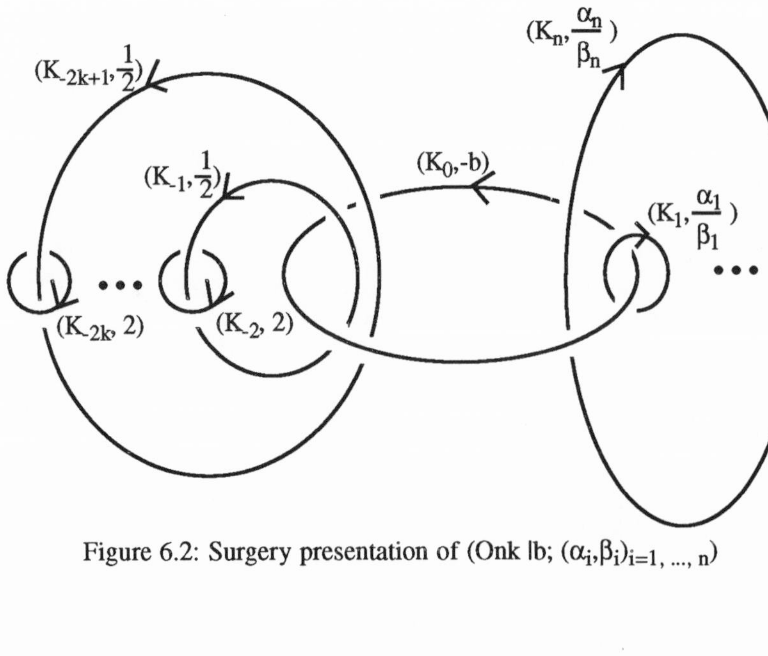

Figure 6.2: Surgery presentation of (Onk |b; $(\alpha_i,\beta_i)_{i=1,\ldots,n}$)

The oriented Seifert fibered space with $\#^k\mathbb{R}P^2$ as its base and with n exceptional fibers $(\text{Onk} \mid b; (\alpha_i,\beta_i)_{i=1,\ldots,n})$ $(\alpha_i, \beta_i \in \mathbb{N}, 0<\beta_i<\alpha_i, \text{g.c.d.}(\alpha_i,\beta_i) = 1)$ is presented by the surgery diagram shown in Figure 6.2.

REMARK: In both cases, the meridian of K_0 is a generic fiber of the Seifert fibration and the core of the surgery performed on K_i, for i = 1, ..., n, is an exceptional fiber.

NOTATION: For all these Seifert fibered spaces, e denotes the "Euler number" of the fibration:

$$e = b + \sum_{i=1}^{n} \frac{\beta_i}{\alpha_i}$$

PROPOSITION **6.1.1**:
The invariant λ *of all oriented Seifert fibered spaces is given by the following formulae.*

For the Seifert fibered spaces with an orientable base:
Let s(e) *denote:*

$$s(e) = \begin{cases} \text{sign}(e) & \text{if } e \neq 0 \\ -1 & \text{if } e = 0 \end{cases}$$

To.0

$$\lambda(\text{Oo0} \mid b; (\alpha_i,\beta_i)_{i=1,\ldots,n}) =$$
$$\left(\frac{s(e)}{24}\left(2 - n + \sum_{i=1}^{n} \frac{1}{\alpha_i^2}\right) + \frac{|e|\, e}{24} - \frac{e}{8} - \frac{|e|}{2}\sum_{i=1}^{n} s(\beta_i,\alpha_i)\right)\prod_{i=1}^{n}\alpha_i$$

To.1

$$\lambda(\text{Oo1} \mid b; (\alpha_i,\beta_i)_{i=1,\ldots,n}) = -\, s(e)\prod_{i=1}^{n}\alpha_i$$

To.>1 If g > 1,

$$\lambda(\text{Oog} \mid b; (\alpha_i,\beta_i)_{i=1,\ldots,n}) = 0$$

For the Seifert fibered spaces with a non-orientable base:
Tn.1

$$\lambda(\text{On1} \mid b; (\alpha_i,\beta_i)_{i=1,\ldots,n}) = 2\left(\frac{e}{8} - \sum_{i=1}^{n} s(\beta_i,\alpha_i)\right)\prod_{i=1}^{n}\alpha_i$$

Tn.2

$$\lambda(\mathrm{On2}\ |b;\ (\alpha_i,\beta_i)_{i=1,\ldots,n}) = \frac{2}{3}\prod_{i=1}^{n}\alpha_i$$

Tn.>2 If $k > 2$,

$$\lambda(\mathrm{Onk}\ |b;\ (\alpha_i,\beta_i)_{i=1,\ldots,n}) = 0$$

6.1.B Proving Proposition 6.1.1 from the formula

(a) *Preliminaries shared by the different cases (simplifications in the surgery formula* 1.7.3*)*
Call $\mathbb{L}$ a surgery presentation as in Figure 6.1 or Figure 6.2 and call A the set indexing its components.

Computing the relative ζ-coefficients involved
• Let K_a be a component of L.
If $n(K_a)$ denotes the number of components of L which algebraically link K_a, then

$$\zeta_A(K_a) = \frac{n(K_a) - 1}{24}$$

because K_a is trivial and the linking number of K_a with any other component of L is 1 or 0.
• Let L_I be a several-component sublink of L ($I \subset A$, $\#I > 1$).
ASSERTION: *If* L_I *is a Borromean link* $(K_0, B_{-2h+1}, B_{-2h})$ *of Figure* 6.1, *then* $\zeta_A(L_I)$ *is equal to* 1; *and* $\zeta_A(L_I)$ *is zero in any other case.*
PROOF OF THE ASSERTION: Since there is no circle with support A, with more than 2 vertices, and with a nonzero associate linking number, we have

$$\zeta_A(L_I) = \zeta(L_I)$$

Furthermore, if L_I is not a Borromean link $(K_0, B_{-2h+1}, B_{-2h})$, then L_I satisfies:

- L_I is split, or
- L_I contains one component together with one of its meridians, or
- $\#I > 3$, and the components of L_I do not link algebraically each other.

So, according to 2.3.7, 2.3.8, and 5.1.6 and 5.2.9,

$$\zeta(L_I) = 0$$

It suffices now to prove that the ζ-coefficient of the Borromean link

$$\mathbb{B} = (K_0, B_{-2h+1}, B_{-2h})$$

is equal to 1.
This can be seen by computing the Alexander polynomial of $\mathbb{B}$, which is

$$\Delta(\mathbb{B}) = (t_1^{1/2} - t_1^{-1/2})(t_2^{1/2} - t_2^{-1/2})(t_3^{1/2} - t_3^{-1/2}),$$

or, this can be deduced from the equalities

$$\lambda((S^1)^3) = 1$$

(according to T5.3) and,

$$\lambda((S^1)^3) = \zeta(\mathfrak{B})$$

(according to T1 and the surgery presentation above for $(S^1)^3 = (\mathrm{Oo}1|0)$).

□

Note now that the coefficient $\det(E(\mathbb{L}_{A\setminus I}))$ of $\zeta(L_I)$, for a Borromean link L_I, is different from zero only if $\mathbb{L}$ presents a Seifert fibered space with a torus as its base.

So, according to Expression 1.7.3 of $\mathbb{F}$, we get 6.1.2:

6.1.2 *When the base space is not a torus,*

$$\lambda(\chi(\mathbb{L})) = \frac{\mathrm{sign}(\mathbb{L})}{24}\left(\prod_{a\in A} q_a\right)\sum_{a\in A}\det(E(\mathbb{L}_{A\setminus\{a\}}))\left(n(K_a) - 1 - \frac{1}{q_a^2}\right)$$

$$+ |H_1(\chi(\mathbb{L});\mathbb{Z})|\left(\frac{\mathrm{signature}(E(\mathbb{L})) - n}{8} - \frac{1}{24}\left(\mathrm{tr}(E(\mathbb{L})) - \sum_{i=1}^{n}\frac{\alpha_i}{\beta_i}\right)\right)$$

$$+ \frac{|H_1(\chi(\mathbb{L});\mathbb{Z})|}{2}\sum_{i=1}^{n}\left(s(\alpha_i,\beta_i) + \frac{1}{4} - \frac{\alpha_i}{12\beta_i}\right)$$

where, according to the Dedekind reciprocity law (3.3.3):

$$s(\alpha_i,\beta_i) + \frac{1}{4} - \frac{\alpha_i}{12\beta_i} = -\, s(\beta_i,\alpha_i) + \frac{1}{12\alpha_i\beta_i} + \frac{\beta_i}{12\alpha_i}$$

(b) *When the base space is a sphere*

$$E(\mathbb{L}) = \begin{pmatrix} -b & 1 & 1 & \cdots & 1 \\ 1 & \frac{\alpha_1}{\beta_1} & 0 & \cdots & 0 \\ 1 & 0 & \frac{\alpha_2}{\beta_2} & \cdots & 0 \\ \cdots & \cdots & \cdots & \cdots & \cdots \\ 1 & 0 & 0 & \cdots & \frac{\alpha_n}{\beta_n} \end{pmatrix}$$

$$\det(E(\mathbb{L})) = -\, e\prod_{i=1}^{n}\frac{\alpha_i}{\beta_i}$$

$$|H_1(\chi(\mathbb{L}))| = |e|\prod_{i=1}^{n}\alpha_i$$

$$\mathrm{sign}(\mathbb{L}) = -\, s(e)$$

If $j \in \{1,\ldots, n\}$,

$$\det(E(\mathbb{L}_{A\setminus\{j\}})) = \frac{\beta_j}{\alpha_j}\left(\frac{\beta_j}{\alpha_j} - e\right)\prod_{i=1}^{n}\frac{\alpha_i}{\beta_i}$$

If $e \neq 0$,

$$\text{signature}(E(\mathbb{L})) - n = -\,\text{sign}(e)$$

It suffices to substitute these values into 6.1.2 to get the result in this case.

❑

(c) *When the base space is an orientable surface with genus* $g \geq 1$

The linking matrix $E(\mathbb{L})$ is the matrix above enlarged with 2g rows and 2g columns of zeros. In particular, its nullity (rank of its kernel) is at least 2; and the right-hand side of 6.1.2 is zero in this case because all its terms contain a minor of $E(\mathbb{L})$ of order at least (#A-1) as a factor.

Now, if the base is a torus, then the contribution

$$\text{sign}(\mathbb{L})\prod_{i=1}^{n}\alpha_i$$

(with $\text{sign}(\mathbb{L}) = -\,s(e)$) of the ζ-coefficient of the Borromean link of the presentation given by Figure 6.1 must be added to the right-hand side of 6.1.2 to get $\lambda(\chi(\mathbb{L}))$.

Thus the formula gives the expected results for all fibered spaces with orientable bases.

❑

(d) *When the base space is non-orientable*

The nullity of $E(\mathbb{L})$ is (k-1), therefore Formula 6.1.2 gives 0 when k is greater than 2.

When k equals 1,

$$E(\mathbb{L}) = \begin{pmatrix} 2 & 1 & 0 & 0 & \dots \\ 1 & \frac{1}{2} & 1 & 0 & \dots \\ 0 & 1 & -b & 1 & \dots \\ 0 & 0 & 1 & \frac{\alpha_1}{\beta_1} & \dots \\ \dots & \dots & \dots & \dots & \dots \end{pmatrix}$$

$$|H_1(\chi(\mathbb{L})| = 4\prod_{i=1}^{n}\alpha_i,\ \det(E(\mathbb{L})) = -2\prod_{i=1}^{n}\frac{\alpha_i}{\beta_i},\ \text{sign}(\mathbb{L}) = -1,$$

$$\text{signature}(E(\mathbb{L})) - n = 1.$$

Again, it suffices to substitute this into 6.1.2 to conclude.

When k is 2, the only nonzero $\det(E(\mathbb{L}_{A\setminus\{a\}}))$ are

$$\det(E(\mathbb{L}_{A\setminus\{-1\}})) = \det(E(\mathbb{L}_{A\setminus\{-3\}})) = -4\prod_{i=1}^{n}\frac{\alpha_i}{\beta_i}$$

and

$$\det(E(\mathbb{L}_{A\setminus\{-2\}})) = \det(E(\mathbb{L}_{A\setminus\{-4\}})) = -\prod_{i=1}^{n}\frac{\alpha_i}{\beta_i}$$

Proposition 6.1.1 is now proved in all cases.

❑

6.1.C Comparison with T5.≥1

Since the H_1 of a Seifert fibered space maps onto the H_1 of its base, it is clear, from T5.>3 that

$$\lambda(\mathrm{Oog}\ |b;\ (\alpha_i,\beta_i)_{i=1,\,\dots,\,n}) = 0 \text{ if } g > 1$$
$$\lambda(\mathrm{Onk}\ |b;\ (\alpha_i,\beta_i)_{i=1,\,\dots,\,n}) = 0 \text{ if } k > 4$$

Let M *be a Seifert fibered space with a non-orientable base.*
The projection from M onto its base space induces an isomorphism on $H_1(.;\mathbb{Q})$ and

$$|\mathrm{Torsion}(H_1(M;\mathbb{Z}))| = 4\prod_{i=1}^{n}\alpha_i$$

This proves together with Proposition 6.1.1, that, *if the base of* M *is* $\mathbb{R}P^2\#\mathbb{R}P^2$, *then:*

$$\Delta''(M)(1) = 2\prod_{i=1}^{n}\alpha_i = \frac{|\mathrm{Torsion}(H_1(M))|}{2}$$

If the base of M *is* $\#^4\mathbb{R}P^2$, that is, the connected sum of a torus and a Klein bottle, then $H_2(M;\mathbb{Z})$ is generated by the regular preimages of "the" two factors S^1 of the torus and a characteristic (for the orientation) curve of the Klein bottle. The intersection of these three surfaces is empty, so T5.3 agrees with the previous computation to confirm that $\lambda(M)$ must be zero.

Now note the following fact which holds for any orientable Seifert fibered space M with a non-orientable base:

FACT **6.1.3:** *A generic fiber* $\mathfrak{F}$ *of the fibration has order* 2 *in* $H_1(M;\mathbb{Z})$ *and does not link algebraically the other fibers.*

PROOF: We can construct a Seifert surface for $2\mathfrak{F}$ as follows:

Take a loop γ in the base of the fibration, based on the projection of $\mathfrak{F}$, such that the neighborhood of γ (in the base) is non-orientable, and the fibration is regular over γ. The preimage of γ under the fibration map is a Klein bottle B. Removing a regular neighborhood of $\mathfrak{F}$ from B yields a cylinder with boundary $2\mathfrak{F}$.

This cylinder is the Seifert surface we were looking for.

❑

If the base space of M *is* $\#^3\mathbb{R}P^2$, that is, the connected sum of a torus and a projective plane, then the regular preimages under the fibration of "the" two factors S^1 of the torus form a basis of $H_2(M;\mathbb{Z})$. They intersect along a generic fiber and give it the same trivialization as the fibration does. So T5.2 and 6.1.3 prove, as the surgery formula did, that $\lambda(M)$ must also be zero in this case.

Now let M *be a Seifert fibered space with a torus as its base.*
If M *has rank* 2, according to T5.2,

$$\frac{-\lambda(M)}{|\mathrm{Torsion}(H_1(M;\mathbb{Z}))|}$$

is the linking number of two generic fibers, which can be directly computed and is equal to 1/e. This is again consistent with the computation of 6.1.B.
If M *has rank* 3
The number e must be zero.
Call q the lowest common multiple of the α_i's, then

$$|\mathrm{Torsion}(H_1(M;\mathbb{Z}))| = \mathrm{g.c.d.}\left\{\frac{1}{\alpha_j\alpha_k}\prod_{i=1}^{n}\alpha_i\right\}_{(j,k)\in\{1,\dots,n\}^2} = \frac{1}{q^2}\prod_{i=1}^{n}\alpha_i$$

We can choose a basis (A,B,C) of $H_2(M;\mathbb{Z})$ such that A and B are preimages under the fibration of two factors S^1 of the torus, and so that (A,B,C) is dual, for the intersection, to a basis (a,b,c) of $H_1(M;\mathbb{Z})$/Torsion where b and a are sections of the two factors S^1. Thus, since a generic fiber $\mathfrak{F}$ intersects neither A nor B, $\mathfrak{F}$ represents $\pm kc$ in $H_1(M;\mathbb{Z})$/Torsion, for an integer k which is the quotient of $|\mathrm{Torsion}(\frac{H_1(M,\mathbb{Z})}{\mathbb{Z}[\mathfrak{F}]})|$ by $|\mathrm{Torsion}(\frac{H_1(M,\mathbb{Z})}{\mathbb{Z}c})| = |\mathrm{Torsion}(H_1(M,\mathbb{Z})|$,
where

$$|\mathrm{Torsion}(\frac{H_1(M,\mathbb{Z})}{\mathbb{Z}[\mathfrak{F}]})| = \mathrm{g.c.d.}\left\{\frac{1}{\alpha_j}\prod_{i=1}^{n}\alpha_i\right\}_{j\in\{1,\dots,n\}} = \frac{1}{q}\prod_{i=1}^{n}\alpha_i$$

So, k = q.
Now, since $\mathfrak{F}$ represents the intersection of A and B,

$$D^{-1}(A)\cup D^{-1}(B)\cup D^{-1}(C)([M])$$

is, up to sign, the intersection number of C and $\mathfrak{F}$, that is k.

$$D^{-1}(A)\cup D^{-1}(B)\cup D^{-1}(C)([M]) = \pm k$$

Thus, again, there is no contradiction.

❑

§6.2 The formula involving the figure-eight linking

6.2.A Proof of the equivalence between Definitions 1.4.8 and 1.7.3 of $\mathbb{F}$

Since

$$- \det(E(\mathbb{L}))\sum_{i\in N}\frac{p_i}{q_i} = (-1)^{\#\emptyset-1}\det(E(\mathbb{L}_{N\setminus\emptyset}))\sum_{i\in N\setminus\emptyset}Lk_c(L_{\emptyset\cup\{i\}})$$

to prove that Definitions 1.4.8 and 1.7.3 of $\mathbb{F}$ are equivalent, it suffices to prove Equality 6.2.1.

6.2.1

$$\sum_{\{I\,/\,I\neq\emptyset,\ I\subset N\}} (-1)^{\#I}\det(E(\mathbb{L}_{N\setminus I}))L_8(\mathbb{L}_I)$$

$$= \sum_{\{J\,/\,J\subset N\}}(-1)^{\#J-1}\det(E(\mathbb{L}_{N\setminus J}))\sum_{i\in N\setminus J}Lk_c(L_{J\cup\{i\}})$$

We prove 6.2.1:

$\det(E(\mathbb{L}_{N\setminus I}))$ is the sum running over the graphs $G(\sigma)$ of permutations σ of $N\setminus I$ of the signature$(\sigma)Lk(\mathbb{L};\sigma)$ (see 2.4.5). The graph of such a permutation is a disjoint union of k oriented circles and its signature is

$$\text{signature}(\sigma) = (-1)^{\#(N\setminus I)-k}$$

So, the left-hand side of 6.2.1 is the sum running over all graphs $G = G(\sigma)\cup G_8$ with support N which are disjoint unions of k oriented circles and a figure-eight graph G_8 of the terms

$$(-1)^{n+k}Lk(\mathbb{L};G)$$

where a figure-eight graph has an upper loop and a lower loop, and each of them has an orientation. (A graph with support I which is homeomorphic to a figure-eight graph appears, in general, eight times in the sum $L_8(\mathbb{L}_I)$.)

With a graph $G = G(\sigma)\cup G_8$ as above, we associate the quadruple $(i,J,\rho,\mathfrak{C})(G)$ where

- i is the element of N corresponding to the crossing point of G_8,
- J is the set of elements of $N\setminus\{i\}$ corresponding to vertices of the upper loop of G_8,
- ρ is the permutation of $N\setminus J$ with the union of $G(\sigma)$ and the lower loop of G_8 as its graph,
- $\mathfrak{C}$ is the upper loop of G_8 which is an oriented circle with support $J\cup\{i\}$.

Now, the right-hand side of 6.2.1 can be seen as the sum running over the $(i,J,\rho,\mathfrak{C})(G)$ of the terms

$$(-1)^{\#J-1}\text{signature}(\rho)\text{Lk}(\mathbb{L};\rho)\text{Lk}(\mathbb{L};\mathfrak{T}) = (-1)^{n+k}\,\text{Lk}(\mathbb{L},G)$$

and we are done.

❑

6.2.B Proof of 1.7.4

Let m_i be the oriented meridian of K_i, and let ν_i be the element of $H_1(\partial T(K_i);\mathbb{Q})$ which represents 0 in $H_1(M\backslash K_i;\mathbb{Q})$ and which satisfies:

$$<m_i,\nu_i> = <m_i,\ell_i> = 1$$

Then the restriction of $\text{Lk}_M(.,K_i)$ on $H_1(\partial T(K_i);\mathbb{Q})$ is $<.,\nu_i>$. In particular,

$$<\mu_i,\nu_i> = p_i$$

$$<\ell_i,\nu_i> \equiv \text{Lk}_M(K_i,K_i) \bmod \mathbb{Z}$$

Now,

$$\ell_i = \nu_i + <\ell_i,\nu_i>m_i$$

So,

$$<\mu_i,\ell_i> = p_i - q_i <\ell_i,\nu_i> \equiv p_i - q_i\,\text{Lk}_M(K_i,K_i) \bmod q_i$$

❑

§6.3 Congruences and relations with the Rohlin invariant

6.3.A Stating the congruences

We will use the surgery formula T1 written in the form:
For any integral surgery presentation $\mathbb{L} = (K_i, \ell_{ii})_{i\in N}$ $(\ell_{ii} \in \mathbb{Z})$ *in* S^3,

6.3.1

$$24\lambda(\chi(\mathbb{L})) - 3|H_1(\chi(\mathbb{L}))|\text{signature}(E(\mathbb{L}))$$

$$=$$

$$\text{sign}(\mathbb{L})\sum_{i\in N}\det(E(\mathbb{L}_{N\backslash\{i\}}))(\,12\Delta''(K_i)(1) - 2 - \ell_{ii}^2\,)$$

$$+\ \text{sign}(\mathbb{L})\sum_{\{I\,/\,I\subset N,\ \#I\geq 2\}}\det(E(\mathbb{L}_{N\backslash I}))(\,24\zeta(L_I) + (-1)^{\#I}L_8(\mathbb{L}_I)\,)$$

(see 1.4.8) to prove the propositions:

PROPOSITION **6.3.2**:
For any closed oriented 3-manifold M, $12\lambda(M)$ *belongs to* $\mathbb{Z}$.

❑

PROPOSITION **6.3.3**:
For any closed oriented 3-manifold M,

$$4\lambda(M) \textit{ belongs to } \mathbb{Z} \textit{ if and only if } H_1(M;\frac{\mathbb{Z}}{3\mathbb{Z}}) \neq \frac{\mathbb{Z}}{3\mathbb{Z}}$$

❑

PROPOSITION **6.3.4**:
Let $\mathbb{L}$ *be an integral surgery presentation in* S^3 *such that* $E(\mathbb{L})$ *has an even diagonal and an odd determinant, then*

$$24\mathbb{F}(\mathbb{L}) - 3\,|\det(E(\mathbb{L}))|\,\mathrm{signature}(E(\mathbb{L})) \in 16\mathbb{Z}$$

❑

These congruences will be proved in Subsection 6.3.B. They immediately prove the invariance of the Rohlin invariant and its Casson-Walker relation to the Casson-Walker invariant; this will be discussed in 6.3.C.

To prove Propositions 6.3.2 to 6.3.4, we will also use the following standard fact :

FACT **6.3.5**
Let $\mathbb{L}$ *be an n-component integral surgery presentation in* S^3, *then any integral symmetric matrix representing the same bilinear symmetric form as* $E(\mathbb{L})$ *over* $\mathbb{Z}^n$ *is the linking matrix of a surgery presentation of* $\chi(\mathbb{L})$.
PROOF: $E(\mathbb{L})$ is the matrix of a bilinear symmetric form ℓ with respect to a basis $(\kappa_1, \ldots, \kappa_n)$.
Note that reversing the orientation of the component K_1 of $\mathbb{L}$ transforms κ_1 into $(-\kappa_1)$ in this basis. Thus, it suffices to prove, for $\{i,j\}\subset\{1,\ldots,n\}$ $(i\neq j)$ and for $\varepsilon=\pm 1$, that the matrix of ℓ with respect to the basis $(\kappa_1, \kappa_2, \ldots, \kappa_{i-1}, \kappa_i + \varepsilon\kappa_j, \ldots, \kappa_n)$ is the linking matrix $E(\mathbb{L}')$ of a surgery presentation $\mathbb{L}'$ of $\chi(\mathbb{L})$. Indeed, such changes of basis generate $SL(n,\mathbb{Z})$.
Since such a $\mathbb{L}'$ can be obtained from $\mathbb{L}$ by a Kirby band move (see [Kir 2] or the description of the move below), we are done.
3-dimensional description of the mentioned Kirby move
Isotope (the trace of) $\mathbb{L}\setminus(K_j,\mu_j)$ in $\chi(K_j,\mu_j)$, by pushing (K_i,μ_i) across the core $C(K_j)$ of the surgery performed on K_j to get $(p(K_i),p(\mu_i))$, so that, in $\chi(K_j,\mu_j)$, $p(K_i)$ is the connected sum of K_i and an ε-oriented meridian of $C(K_j)$, and the other components of $\mathbb{L}$ have not moved.
The union $\mathbb{L}'$ of $\mathbb{L}\setminus(K_i,\mu_i)$ with the trace in S^3 of $(p(K_i),p(\mu_i))$ is clearly another presentation of $\chi(\mathbb{L})$. The trace of $p(K_i)$ is a band connected sum of K_i with $\varepsilon\mu_j$.

❑

6.3.B Proving the congruences

We need:

LEMMA **6.3.6**

(A) *If* K *is a knot in* S^3, *then*

$$\Delta''(K)(1) \in 2\mathbb{Z}$$

(B) *Let* L *be a link with* n *components* $K_1, \ldots, K_n$, $n \geq 2$, *in* S^3, *then*

$$\zeta(L) \in \frac{1}{4}\mathbb{Z}$$

(B.a) *If for any* $i \in N = \{1, \ldots, n\}$, *either* $\sum_{j \in N\setminus\{i\}} Lk(K_j, K_i)$ *is odd, or* $Lk(K_i, K_j)$ *is even for any* $j \in N\setminus\{i\}$, *then*

$$\zeta(L) \in \begin{cases} \mathbb{Z} & \text{if } n > 2 \\ \frac{1}{2}\mathbb{Z} & \text{if } n = 2 \end{cases}$$

(B.b) *If for any* $i \in N$, $\sum_{j \in N\setminus\{i\}} Lk(K_j, K_i)$ *is odd, then*

$$\zeta(L) \in 2\mathbb{Z}$$

PROOF:

(A) $\Delta(K) \in \mathbb{Z}[t, t^{-1}]$. (See 2.3.13.)

(B) Let I be the set of the elements i of N such that $\sum_{j \in N\setminus\{i\}} Lk(K_j, K_i)$ is even.

Then, according to 2.3.9,

$$\mathcal{D}(L) \prod_{i \in I} \exp(\tfrac{u_i}{2}) \in \mathbb{Z}[\exp(\pm u_i)]_{i=1, \ldots, n}$$

So, according to 2.5.1 and 2.5.2 (see also 2.4.1 for the notations),

$$\frac{\partial^n}{\partial u_1 \ldots \partial u_n} (\mathcal{D}(L) \prod_{i \in I} \exp(\tfrac{u_i}{2}))(0) = \mathcal{D}_{\mathbb{1}}(L) + \sum_{\{i,j\} \subset I} \frac{\mathcal{D}_{\mathbb{1}ij}(L)}{4} \quad \in \mathbb{Z}$$

this proves (B) since the $\mathcal{D}_{\mathbb{1}ij}(L)$ are integers. (See 2.4.3, 2.4.4 and 2.5.2 for $\mathcal{D}_{\mathbb{1}ij}(L)$.)

(B.a) If n=2, then either $I = \emptyset$, or $I = \{1,2\}$ and $\mathcal{D}_{\mathbb{1}12}(L) = -Lk(K_1, K_2)$ is even. If n>2 and if $\{i,j\} \subset I$, then the $\mathbb{1}_{ij}$-trees, which are segments with ends i and j, contain at least two edges with even linking numbers (the two different edges containing i and j). So, 4 divides $\mathcal{D}_{\mathbb{1}ij}(L)$.

(B.b) In this case, $\Delta(L) \in \mathbb{Z}[t_i^{\pm 1}]_{i=1, \ldots, n}$ and

$$\Delta(L)(t_1, \ldots, t_n) = (-1)^n \Delta(L)(t_1^{-1}, \ldots, t_n^{-1}).$$

❑

PROOF OF PROPOSITION 6.3.2

First note that the right-hand side of the surgery formula 6.3.1 is always an integer.

The sum $L_8(\mathbb{L}_I)$ is even if #I is bigger than 1 because of the symmetry exchanging the upper loop and the lower one in a figure-eight graph.

Using Fact 6.3.5 to choose a surgery presentation $\mathbb{L}$ for M with an appropriate linking matrix modulo 2, we see easily that

$$3|\det(E(\mathbb{L}))|\text{signature}(E(\mathbb{L})) - \text{sign}(\mathbb{L})\sum_{i\in N}\det(E(\mathbb{L}_{N\setminus\{i\}}))\ell_{ii}^2$$

is even. Thus, the same can be said of $24\lambda(\chi(\mathbb{L})) = 24\lambda(M)$.

❑

PROOF OF PROPOSITION 6.3.3

Use Fact 6.3.5 to choose a surgery presentation $\mathbb{L}$ of M with a (mod 3)-diagonal linking matrix (by diagonalizing the linking matrix of a surgery presentation of M over $\mathbb{Z}/3\mathbb{Z}$). Then, according to 6.3.1,

$$24\lambda(\chi(\mathbb{L})) \equiv_3 - \text{sign}(\mathbb{L})\sum_{i\in N}\det(E(\mathbb{L}_{N\setminus\{i\}}))(\,2 + \ell_{ii}^2\,)$$

The ith term of this sum is nonzero mod 3 if and only if ℓ_{ii} belongs to $3\mathbb{Z}$ and $\det(E(\mathbb{L}_{N\setminus\{i\}}))$ does not. This may happen only when there is exactly one i, say i_0, such that 3 divides ℓ_{ii}, that is, only when $\dim(H_1(M;\mathbb{Z}/3\mathbb{Z})) = 1$, and, in this case:

$$24\lambda(\chi(\mathbb{L})) \equiv_3 \text{sign}(\mathbb{L})\det(E(\mathbb{L}_{N\setminus\{i_0\}})) \not\equiv_3 0$$

So, $4\lambda(M) \notin \mathbb{Z}$.

❑

PROOF OF PROPOSITION 6.3.4

Our hypotheses imply that $E(\mathbb{L})$ is congruent (mod 2) to a block diagonal matrix consisting of blocks of the type $\begin{pmatrix} 0 & 1 \\ 1 & 0 \end{pmatrix}$. Thus, according to Fact 6.3.5, we may transform $\mathbb{L}$ without changing signature$(E(\mathbb{L}))$ nor $\chi(\mathbb{L})$, so that $E(\mathbb{L})$ has this form. So we may decompose the set of indices of the components of $\mathbb{L}$ as $N = K \cup a(K)$, where K and a(K) are disjoint, and a is an involution of N so that:

$$\forall\ (i,j) \in N^2,\ \ell_{ij} \equiv_2 \begin{cases} 0 & \text{if } j \neq a(i) \\ 1 & \text{if } j = a(i) \end{cases}$$

Now, according to T1, it suffices to prove that

$$24\mathbb{F}(\mathbb{L}) - 3\,|\det(E(\mathbb{L}))|\,\mathrm{signature}(E(\mathbb{L}))$$
$$= \mathrm{sign}(\mathbb{L})\sum_{i\in N}\det(E(\mathbb{L}_{N\setminus\{i\}}))(\,12\Delta''(K_i)(1) - 2 - \ell_{ii}^2\,)$$
$$+\ \mathrm{sign}(\mathbb{L}) \sum_{\{I\,/\,I\subset N,\ \#I\geq 2\}}\det(E(\mathbb{L}_{N\setminus I}))(\,24\zeta(L_I) + (-1)^{\#I}L_8(\mathbb{L}_I)\,)$$

is multiple of 16 under these further assumptions.

For any subset I of N, $\det(E(\mathbb{L}_{N\setminus I}))$ is multiple of $2^{\#\mathfrak{Z}(I)}$, where $\mathfrak{Z}(I)$ denotes the set of the elements i of I such that $a(i)\notin I$.

Lemma 6.3.6 proves in particular:

$$\zeta(L_I) \in 2^{1-\#\mathfrak{Z}(I)}\mathbb{Z} \qquad \text{if } \#I \geq 2$$

So the terms involving the $\Delta''(K_i)(1)$ and the $\zeta(L_I)$ are zero (mod 16).

Let I be a subset of N, $\#I \geq 2$, we now prove that $\det(E(\mathbb{L}_{N\setminus I}))L_8(\mathbb{L}_I)$ is zero (mod 16) unless I is a pair of associate elements (under the involution a).
Let G_8 be a figure-eight graph with support I.
Since #I is larger than one, the two loops of G_8 are not isomorphic. So, if $s(G^8)$ is the number of loops of G_8 which have at least 3 vertices (those which change under an orientation reversal), exactly $2^{s(G_8)+1}$ graphs isomorphic to G_8 appear in $L_8(\mathbb{L}_I)$.
If $\#I \geq 4$, then $s(G_8) \geq 1$ and G_8 has at least two even edges (i.e., edges with an even associate linking number). So, in this case, 16 divides $L_8(\mathbb{L}_I)$.
If $\#I = 3$, then G_8 has at least 2 even edges, and $\det(E(\mathbb{L}_{N\setminus I}))$ must be even. So, 16 divides $\det(E(\mathbb{L}_{N\setminus I}))L_8(\mathbb{L}_I)$.
If I is a pair of non-associate elements, then 16 divides again $L_8(\mathbb{L}_I)$.

Observe also that:

$$\det(E(\mathbb{L}_{N\setminus\{i\}})) \equiv_8 \ell_{a(i)a(i)}\det(E(\mathbb{L}_{N\setminus\{i,a(i)\}}))$$

All these cancellations reduce Formula 6.3.1 mod 16 to

$$24\mathbb{F}(\mathbb{L}) - 3\,|\det(E(\mathbb{L}))|\,\mathrm{signature}(E(\mathbb{L}))$$
$$\equiv_{16} -\,\mathrm{sign}(\mathbb{L})\sum_{i\in K}\det(E(\mathbb{L}_{N\setminus\{i,a(i)\}}))(\,\ell_{a(i)a(i)}(\,2 + \ell_{ii}^2\,) + \ell_{ii}(\,2 + \ell_{a(i)a(i)}^2\,)\,)$$
$$+\ \mathrm{sign}(\mathbb{L})\sum_{i\in K}\det(E(\mathbb{L}_{N\setminus\{i,a(i)\}}))(\,2\ell_{a(i)a(i)}\ell_{ia(i)}^2 + 2\ell_{ii}\ell_{ia(i)}^2\,)$$

$$\equiv_{16} 2\,\mathrm{sign}(\mathbb{L}) \sum_{i\in K} \det(E(\mathbb{L}_{N\setminus\{i,a(i)\}}))\ell_{a(i)a(i)}(\ell^2_{ia(i)}-1)$$

$$+\, 2\,\mathrm{sign}(\mathbb{L}) \sum_{i\in K} \det(E(\mathbb{L}_{N\setminus\{i,a(i)\}}))\ell_{ii}(\ell^2_{ia(i)}-1)$$

$$-\,\mathrm{sign}(\mathbb{L}) \sum_{i\in K} \det(E(\mathbb{L}_{N\setminus\{i,a(i)\}}))\ell_{ii}\ell_{a(i)a(i)}(\ell_{ii}+\ell_{a(i)a(i)})$$

and these three summands are multiples of 16.

❑

6.3.C A short trip in the fourth dimension, relating λ to the signature of spin manifolds

A few basic things about spin structures

(The reader is referred to [Mil], [G-M 1] or [Kir 1] for more details.)

Let ξ be a principle fibre bundle with structural group SO(n) ($n\geq 3$), total space $E(\xi)$ and base space $B(\xi)$.

Define a *spin structure* on ξ as a cohomology class σ in $H^1(E(\xi); \mathbb{Z}/2\mathbb{Z})$ whose restriction to each fibre is the generator of $H^1(\text{Fibre}; \mathbb{Z}/2\mathbb{Z}) \cong \mathbb{Z}/2\mathbb{Z}$.

Such a spin structure σ provides a natural homotopy class of trivializations of ξ over any loop λ of $B(\xi)$: the class of the loops lifting λ which are sent to zero by σ.

When $B(\xi)$ is a CW-complex, we get thus a *second (equivalent) definition of a spin structure* on ξ: A *spin structure* on ξ is a homotopy class of trivializations of ξ over the one-skeleton of $B(\xi)$ which extend over the two-skeleton of $B(\xi)$.

The obstruction $w_2(\xi)\in H^2(B(\xi); \mathbb{Z}/2\mathbb{Z})$ to the existence of spin structures can then be seen as the obstruction to trivializing (the pull-back of) ξ over surfaces mapped to $B(\xi)$.

The set of spin structures can be seen from both definitions as an affine space with associated vector space $H^1(B(\xi); \mathbb{Z}/2\mathbb{Z})$.

A spin structure on an oriented Riemannian n-manifold M ($n\geq 3$) is a spin structure on the principle bundle $\xi(M)$ of oriented orthonormal n-frames on M. M is said to be spin if it is equipped with a spin structure.

If $n\geq 4$ and if $\partial M\neq\emptyset$, then restricting a spin structure on M

$$\sigma\in H^1(E(\xi(M)); \frac{\mathbb{Z}}{2\mathbb{Z}})$$

to $H^1(E(\xi(\partial M)); \mathbb{Z}/2\mathbb{Z})$ yields a natural spin structure on ∂M (the inclusion from SO(n-1) to SO(n) induces an isomorphism on $H^1(.; \mathbb{Z}/2\mathbb{Z})$).

If M is an oriented Riemannian 4-manifold, and if F is a closed surface immersed in M, then ξ has a trivialization over F if and only if the (mod 2) self-intersection of F in M is zero (see [G-M 1], p.43). So, M has a spin structure if and only if the intersection form maps the diagonal of $H_2(M;\mathbb{Z}/2\mathbb{Z})^2$ to zero.

Let $\mathbb{L} = (K_i,\mu_i)_{i=1,\dots,n}$ be an n-component integral surgery presentation in S^3. View S^3 as the boundary of the ball B^4.
We denote by $W_{\mathbb{L}}$ the simply connected 4-manifold obtained by adding to B^4 two-handles $(D^2{\times}D^2)_i$, identifying $(\partial D^2 \times D^2)_i$ with $K_i \times D^2 \subset S^3 \subset B^4$ so that $(\partial D^2 \times \{1\})_i$ is identified with μ_i, and by smoothing in a standard way.
The boundary of $W_{\mathbb{L}}$ is $\chi(\mathbb{L})$; and $E(\mathbb{L})$ is the matrix of the intersection form on $W_{\mathbb{L}}$ with respect to the basis of $H_2(W_{\mathbb{L}};\mathbb{Z})$ associated with the handle decomposition of $W_{\mathbb{L}}$ mentioned above. So, $W_{\mathbb{L}}$ has a spin structure if and only if the diagonal of $E(\mathbb{L})$ is zero modulo 2.

We can now use Proposition 6.3.4 to prove the well-known:

THEOREM (Rohlin):
The signature of a smooth closed spin 4-manifold is divisible by 16.

PROOF: Let M_1 be such a (connected) manifold, equip it with a Morse function f_1 with only one minimum and only one maximum (and critical points with other indices). Add handles $(D^2{\times}D^3)$ to the (naturally) spin 5-manifold $M_1 \times I$ so that it becomes a spin cobordism between M_1 and a connected spin 4-manifold M which can be equipped with a Morse function without critical points of index 1 or 3.
(Gluing a $(D^2{\times}D^3)$ allows us to replace an $S^1{\times}D^3$ in M_1 corresponding to an index 1 (or 3) critical point of f_1 by a $D^2{\times}S^2$. Such a $(D^2{\times}D^3)$ can be glued so that the spin structure extends to $D^2{\times}D^3$ as can be seen from the second given definition of a spin structure.)
Since M and M_1 are cobordant, they have the same signature.
Now, M minus a 4-ball B^4 is the $W_{\mathbb{L}}$ of a surgery presentation $\mathbb{L}$ of S^3. According to T1,

$$\mathbb{F}(\mathbb{L}) = 0$$

So, according to 6.3.4, the signature of $E(\mathbb{L})$, which is the signature of M, is divisible by 16.

❑

Recall the following:

THEOREM (Milnor): *Any closed spin 3-manifold bounds a smooth compact spin 4-manifold.*

❑

CONSEQUENCE OF THE ROHLIN THEOREM:

Let M be a $\frac{\mathbb{Z}}{2\mathbb{Z}}$-sphere, the signature $\sigma(M)$ of a smooth compact spin 4-manifold with boundary M is an invariant of M mod 16. The invariant

$$\sigma(M) \in \frac{\mathbb{Z}}{16\mathbb{Z}}$$

is called the *Rohlin invariant* of M.

PROOF: Since M has only one spin-structure, if W and W' are two smooth spin 4-manifolds with boundary M, then $(W \cup_M -W')$ is spin, and its signature is divisible by 16 because of the Rohlin theorem. So, the signatures of W and W' are congruent mod 16.

❑

FACT **6.3.7**

Any oriented closed 3-manifold has an integral surgery presentation in S^3 *with even surgery coefficients.*

This fact is proved in this form in [Ka] §3, and it proves the Milnor theorem above.

It can also be proved from the theorem of Milnor, mentioned above, using the same arguments as in the proof of the Rohlin theorem above.

❑

PROPOSITION **6.3.8** (Walker, Casson for the $\mathbb{Z}$-spheres): *Comparing* λ *with the Rohlin invariant*

For any $\frac{\mathbb{Z}}{2\mathbb{Z}}$*-sphere* M, $8|H_1(M;\mathbb{Z})|\lambda(M)$ *is congruent to* $\sigma(M)$ *modulo 16.*

PROOF: According to Fact 6.3.7, M has an integral surgery presentation $\mathbb{L}$ in S^3 such that $E(\mathbb{L})$ has an even diagonal. So,

$$\sigma(M) = \text{signature}(E(\mathbb{L})) \bmod 16$$

Since $H_1(M;\mathbb{Z}/2\mathbb{Z})$ is trivial, $\det(E(\mathbb{L}))$ is odd, and as we noted in the proof of Proposition 6.3.4, this implies that $E(\mathbb{L})$ has even rank and thus even signature. Further, Proposition 6.3.4 proves that

$$24\lambda(M) - 3\,|\det(E(\mathbb{L}))|\,\text{signature}(E(\mathbb{L})) \in 16\mathbb{Z}$$

According to 6.3.2 and 6.3.3, this implies

$$8|H_1(M;\mathbb{Z})|\lambda(M) - |H_1(M;\mathbb{Z})|^2\text{signature}(E(\mathbb{L})) \in 16\mathbb{Z}$$

So, since $|H_1(M;\mathbb{Z})|^2$ is congruent to 1 mod 8 (and signature($E(\mathbb{L})$) is even), we are done.

❑

§6.4 The surgery formula in terms of one-variable Alexander polynomials

§6.4 is devoted to the proof of Proposition 1.7.8.

PROPOSITION **1.7.8**
Let $\mathbb{L}$ *be a surgery presentation in a rational homology sphere* M *such that the components of the underlying link of* $\mathbb{L}$ *are null-homologous.*
Then, with the notation 1.7.5 *to* 1.7.7, $\mathbb{F}_M(\mathbb{L})$ *can be written as*

$$\mathbb{F}_M(\mathbb{L}) = \mathrm{sign}(\mathbb{L})\,|H_1(M)|\left(\prod_{i=1}^{n} q_i\right)\sum_{\{J\,/\,J\neq\emptyset\,,\,J\subset N\}} \det(E(\mathbb{L}_{N\setminus J};J))\, a_1(L_J)$$

$$+\ \mathrm{sign}(\mathbb{L})\,|H_1(M)|\left(\prod_{i=1}^{n} q_i\right)\sum_{\{J\,/\,J\neq\emptyset\,,\,J\subset N\}} \frac{\det(E(\mathbb{L}_{N\setminus J}))\,(-1)^{\#J}\,\theta(\mathbb{L}_J)}{24}$$

$$+\ |H_1(\chi_M(\mathbb{L}))|\left(\frac{\mathrm{signature}(E(\mathbb{L}))}{8} + \sum_{i=1}^{n}\frac{s(p_i - q_i Lk_M(K_i,K_i),q_i)}{2}\right)$$

According to 1.4.8, to prove Proposition 1.7.8, it suffices to prove Equality 6.4.1.
6.4.1

$$\sum_{\{I\,/\,I\neq\emptyset,\,I\subset N\}} \det(E(\mathbb{L}_{N\setminus I}))\left(\frac{\zeta(L_I)}{|H_1(M)|} + \frac{(-1)^{\#I}L_8(\mathbb{L}_I)}{24}\right) - \sum_{i\in N}\frac{\det(E(\mathbb{L}_{N\setminus\{i\}}))}{24\,q_i^2}$$

$$= \sum_{\{J\,/\,J\neq\emptyset,\,J\subset N\}}\left(\det(E(\mathbb{L}_{N\setminus J};J))\,a_1(L_J) + \frac{\det(E(\mathbb{L}_{N\setminus J}))\,(-1)^{\#J}\,\theta(\mathbb{L}_J)}{24}\right)$$

To prove 6.4.1, we introduce the notation:
6.4.2

$$\Theta(L_J) = \sum_{\{K\,/\,K\subset J,\,K\neq\emptyset\}} Lk(L;J\setminus K,\rightarrow K)\big(\Theta_b(\mathbb{L}_K)-L_8(\mathbb{L}_K)\big)$$

(See 1.4.6, 1.7.5 and 2.4.6.)

This Θ can be thought of as a sum running over graphs shown in Figure 6.3 of the associated linking numbers.

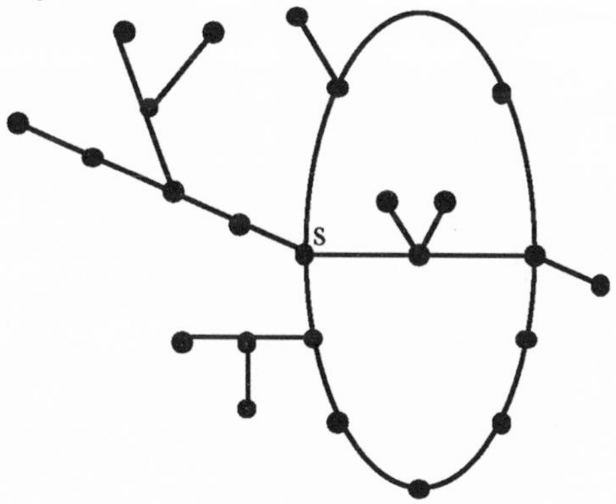

Figure 6.3 : Typical graph involved in Θ : A (possibly) bearded θ

Note that:

6.4.3

$$\Theta(L) = \sum_{\{(I,s)\,/\,I\subset N,\ I\neq\emptyset,\ s\notin I\}} Lk(L;N\setminus I,\rightarrow I)Lk_c(I\cup\{s\})$$

Then, we define $\eta(L_J)$, for any subset J of N, using the notation of 2.3.15.

6.4.4

$$\eta(L_J) = a_1(L_J) + \frac{(\#J-2)a_0(L_J)}{24} + \frac{(-1)^{\#J}\Theta(L_J)}{24}$$

Equality 6.4.1, and hence Proposition 1.7.8, are now the direct consequence of Equalities 6.4.4 to 6.4.8.

6.4.5

$$\frac{\zeta(L_I)}{|H_1(M)|} = \sum_{\{J\,/\,J\neq\emptyset,\ J\subset I\}} \eta(L_J) \sum_{g\in\mathfrak{F}(I\setminus J;J)} Lk(L;g)$$

where $\mathfrak{F}(I\setminus J;J)$ denotes the set of functions from $I\setminus J$ to J.

6.4.6

$$\sum_{\{I,J\,/\,J\neq\emptyset,\ J\subset I\subset N\}} \det(E(\mathbb{L}_{N\setminus I})) \sum_{g\in\mathfrak{F}(I\setminus J;J)} Lk(L;g)\,(-1)^{\#J}\Theta(L_J)$$

$$= \sum_{\{K\,/\,K\neq\emptyset,\ K\subset N\}} \det(E(\mathbb{L}_{N\setminus K}))(-1)^{\#K}(\Theta_b-L_8)(\mathbb{L}_K)$$

6.4.7

$$\sum_{\{ I,J \,/\, J\neq\emptyset,\, J\subset I\subset N \}} \det(E(\mathbb{L}_{N\setminus I})) \sum_{g\in \mathcal{F}(I\setminus J;J)} Lk(L;g)\,(\#J-2)a_0(L_J)$$

$$= -\sum_{i\in N}\det(E(\mathbb{L}_{N\setminus\{i\}})) - 2\sum_{\{i,j\}\subset N} \ell_{ij}\det(E(\mathbb{L}_{N\setminus\{i,j\}}))$$

6.4.8

$$\sum_{\{ I \,/\, J\subset I\subset N \}} \det(E(\mathbb{L}_{N\setminus I})) \sum_{g\in \mathcal{F}(I\setminus J;J)} Lk(L;g) = \det(E(\mathbb{L}_{N\setminus J};J))$$

Equality 6.4.8 is easy, it remains only to prove 6.4.5, 6.4.6 and 6.4.7.

❑

PROOF OF 6.4.5

We first prove Equality 6.4.9 for any link L with indexing set N. ($\#N = n$)

$$\textbf{6.4.9} \qquad |H_1(M)|\eta(L) = \sum_{\{I \,/\, I\neq\emptyset,\, I\subset N\}}\Big((-1)^{n-\#I}Lk(L;N\setminus I, \rightarrow I)\zeta(L_I)\Big)$$

If $n = 1$, then 6.4.9 becomes "$\zeta(L)=|H_1(M)|\eta(L)$" and is true. Assume now $n\geq 2$.
According to 2.5.2 and 6.4.3, the coefficient of u^n in $\mathcal{D}(L)(u, u, \ldots, u)$ is:

$$\sum_{\{R \,/\, |R|=n\}}\mathcal{D}_R(L) = \sum_{\{I \,/\, I\neq\emptyset,\, I\subset N\}}\Big((-1)^{n-\#I}Lk(L;N\setminus I, \rightarrow I)\zeta(L_I)\Big) + \frac{(-1)^{n-1}|H_1(M)|\Theta(L)}{24}$$

According to 2.3.13, 2.3.14 and 2.3.15,

$$\frac{\mathcal{D}(L)(u, u, \ldots, u)}{|H_1(M)|} =$$

$$\left(\exp(\tfrac{u}{2}) - \exp(-\tfrac{u}{2})\right)^{n-2}\left(a_0(L) + a_1(L)\left(\exp(\tfrac{u}{2}) - \exp(-\tfrac{u}{2})\right)^2 + O(u^4)\right)$$

and the coefficient of u^n in $\mathcal{D}(L)(u, u, \ldots, u)$ is thus

$$|H_1(M)|\left(\frac{n-2}{24}a_0(L) + a_1(L)\right)$$

Comparing the two expressions of this coefficient proves 6.4.9.
It suffices to prove 6.4.5 when $L_I = L$, and so we shall induct on $n = \#N$.
If 6.4.5 is true for any $I\subset N$ with $0<\#I<n$, then 6.4.9 becomes:

$$\frac{\zeta(L)}{|H_1(M)|} = \eta(L) + \sum_{\{I \,/\, I\neq\emptyset,\, I\neq N,\, I\subset N\}} \left((-1)^{n-\#I+1} Lk(L;N\backslash I, \to I) \sum_{\{ J \,/\, J\neq\emptyset \,,\, J\subset I \}} \eta(L_J) \sum_{g \in \mathfrak{F}(I\backslash J;J)} Lk(L;g) \right)$$

where (if $J \neq \emptyset, N$)

$$\sum_{\{I \,/\, I\neq N,\, I\subset N,\, J\subset I\}} \left((-1)^{n - \#I + 1} Lk(L;N\backslash I, \to I) \sum_{g \in \mathfrak{F}(I\backslash J;J)} Lk(L;g) \right)$$

$$= (-1)^{n - \#J + 1} \sum_{g \in \mathfrak{F}(N\backslash J;\to J)} Lk(L;g) \sum_{\{B \,/\, B\neq N\backslash J,\, B\subset g^{-1}(J)\ [B=I\backslash J]\}} (-1)^{\#B}$$

$$= \sum_{g \in \mathfrak{F}(N\backslash J,J)} Lk(L;g)$$

because

$$\sum_{\{B \,/\, B\neq N\backslash J,\, B\subset g^{-1}(J)\}} (-1)^{\#B} = \begin{cases} 0 & \text{if } N\backslash J \neq g^{-1}(J) \\ (-1)^{n - \#J + 1} & \text{otherwise} \end{cases}$$

This proves 6.4.5.

❑

PROOF OF 6.4.6

It suffices to prove the equality between the coefficients of the underlying beardless θ's with support K (see Figure 1.2), that is the following equality:

$$\sum_{\{ I,J \,/\, K\subset J\subset I\subset N \}} \det(E(\mathbb{L}_{N\backslash I})) \sum_{g \in \mathfrak{F}(I\backslash J,J)} Lk(L;g)\, (-1)^{\#J} Lk(L;J\backslash K,\to K)$$

$$= (-1)^{\#K} \det(E(\mathbb{L}_{N\backslash K}))$$

where

$$\sum_{\{ J \,/\, K\subset J\subset I \}} \sum_{g \in \mathfrak{F}(I\backslash J,J)} Lk(L;g)\, (-1)^{\#J} Lk(L;J\backslash K,\to K)$$

$$= (-1)^{\#I} \sum_{g \in \mathfrak{F}(I\backslash K,\to K)} Lk(L;g) \sum_{\{B \,/\, B\subset I\backslash (K\cup g(I\backslash K))\ [B=I\backslash J]\}} (-1)^{\#B}$$

and the sum running over the sets B is 0 unless $I \setminus (K \cup g(I \setminus K))$ is empty, that is, unless K is equal to I. This proves 6.4.6.

❑

PROOF OF 6.4.7

Use

$$a_0(L_J) = (-1)^{\#J-1} \sum_{\{T\ /\ T \text{ is a tree with support } J\}} Lk(L;T)$$

(see 2.5.2, 2.3.13, 2.3.14 and 2.3.15) to write

$$\sum_{\{J\ /\ J \neq \emptyset\ ,\ J \subset I\}} \ \sum_{g \in \mathfrak{F}(I \setminus J;J)} Lk(L;g)\ (\#J-2) a_0(L_J) =$$

$$(-1)^{\#I} \sum_{\{T\ /\ T \text{ is a tree with support } I\}} Lk(L;T) \sum_{\substack{\{B\ /\ B \subset \{\text{vertices of T with valence } 1\} \subset I, \\ B \neq I,\ [B = I \setminus J]\}}} (\#I-\#B-2)(-1)^{\#B-1}$$

If #I>2, then any tree T with support I has at least one vertex with valence larger than 1,

$$\sum_{B \subset \{\text{vertices of T with valence } 1\}} (\#I-2)(-1)^{\#B} = 0,$$

and

$$\sum_{B \subset \{\text{vertices of T with valence } 1\}} (\#B)(-1)^{\#B-1}$$

is also zero because it is the derivative at (-1) of $(x+1)^f$ where f is the number of vertices of T with valence 1 which has to be at least 2.

This allows us to consider only the sets I with cardinality 1 and 2 in the left-hand side of 6.4.7 which is clear now.

❑

Appendix

More about the Alexander series

§A.1 Introduction

This appendix proves the assertions of §2.2 and §2.3. It can hopefully be read (along with Chapter 2) as an introduction to Alexander polynomials.

DEFINITIONS **A.1.1**: *Group rings and associated definitions*
If G is a group, $\mathbb{Z}[G]$ denotes the *group ring* of G: $\mathbb{Z}[G]$ is the $\mathbb{Z}$-module freely generated by the elements g of G denoted by e^g or exp(g) (formal notation) when considered as elements of $\mathbb{Z}[G]$; and $\mathbb{Z}[G]$ is equipped with the $\mathbb{Z}$-bilinear multiplication which sends (exp(g),exp(g')) to exp(gg') (or to exp(g+g') if G is commutative and if its group law is written additively).
The linear morphism from $\mathbb{Z}[G]$ to $\mathbb{Z}$ which maps exp(g) to 1, for $g \in G$, is denoted by ε and is called the *augmentation morphism* of $\mathbb{Z}[G]$.
If G is abelian, $\mathbb{Z}[G]$ is also equipped with a natural ring involution which maps exp(g) to exp(-g). This involution is called *conjugation,* and is denoted by ι.
Note that, if G is a free abelian group of rank k, $\mathbb{Z}[G]$ is isomorphic to the polynomial ring $\mathbb{Z}[t_i^{\pm 1}]_{i=1, \ldots, k}$ and is thus a factorial (or unique factorisation) ring.
(The factoriality of $\mathbb{Z}[t_i^{\pm 1}]_{i=1, \ldots, k}$ can easily be deduced from the factoriality of $\mathbb{Z}[t_i]_{i=1, \ldots, k}$ by carefully chosen multiplications by the units of $\mathbb{Z}[t_i^{\pm 1}]_{i=1, \ldots, k}$, which are the monomials with coefficients ± 1.)
The units of $\mathbb{Z}[G]$ of the form exp(g), $g \in G$, are called *positive units.*

A.1.A From the Reidemeister torsion to the Alexander series

In §A.2 and §A.3, we define the (normalized) Reidemeister torsion $\tau(N,o(N))$ of a link exterior N equipped with an orientation $o(N)$ of $(H_1(N;\mathbb{R})\oplus H_2(N;\mathbb{R}))$ and we show:

The torsion $\tau(N,o(N))$ *is an invariant belonging to the field of fractions of*

$$\mathbb{Z}[\frac{1}{2}Q_{fa}(N)=\frac{1}{2}\frac{H_1(N;\mathbb{Z})}{\text{Torsion}}],$$

(where $\frac{1}{2}Q_{fa}(N)$ denotes $\left(\frac{1}{2}\mathbb{Z}\right)\otimes_{\mathbb{Z}} Q_{fa}(N)$) *which satisfies:*

A.1.2

$$\text{If } Q_{fa}(N) \not\cong \mathbb{Z}, \quad \tau(N,o(N)) \in \mathbb{Z}[\frac{1}{2}Q_{fa}(N)]$$

$$\text{If } Q_{fa}(N) = \mathbb{Z}g, \quad \tau(N,o(N))(e^g - 1) \in \mathbb{Z}[\frac{1}{2}Q_{fa}(N)]$$

A.1.3

$$\tau(N, -o(N)) = -\tau(N,o(N))$$

(Before forgetting Alexander series until the end of §A.3,) we assume, in this subsection A.1.A, that the Reidemeister torsion is well-defined and satisfies A.1.2 and A.1.3; and we define the Alexander series (and, according to 2.1.1, the normalized several-variable Alexander polynomial) from this Reidemeister torsion.

From a link exterior to a link

If we suppose that N is the exterior of an oriented n-component link $L = \{K_1, \dots, K_n\}$ in an oriented rational homology sphere M, we have the following morphism (which extends naturally to fields of fractions)

A.1.4

$$\begin{array}{llll} \psi_L: & \mathbb{Z}[\frac{1}{2}Q_{fa}(M\setminus L)] & \to & \mathbb{Z}[\exp(\pm\frac{u_i}{2O_M(K_i)})]_{i=1,\dots,n} \\ & \exp(\frac{1}{2}x) & \to & \exp(\frac{1}{2}\sum_{i=1}^{n} Lk_M(x,K_i)u_i) \end{array}$$

A.1.5 Furthermore, orientations for L and M give $(H_1(N;\mathbb{R})\oplus H_2(N;\mathbb{R}))$ the orientation $o_L(N)$ induced by the basis $(m_1, m_2, \dots, m_n)$ of $H_1(N;\mathbb{R})$ followed by the basis $(\partial T(K_1), \partial T(K_2), \dots, \partial T(K_{n-1}))$ of $H_2(N;\mathbb{R})$, where m_i and $\partial T(K_i)$ denote the oriented meridian of K_i and the oriented boundary of the tubular neighborhood of K_i, respectively.

Note that this orientation $o_L(N)$ is independent of the order of the components.

DEFINITION **A.1.6**: The *Alexander series* of L in M, $\mathcal{D}(L)$, is defined by:

- If $n \geq 2$, $\mathcal{D}(L) = \psi_L(\tau(M\backslash L, o_L(M\backslash L)))$
- If $n = 1$,

$$\mathcal{D}(L) = \psi_L(\tau(M\backslash L, o_L(M\backslash L)))\left(\exp(\frac{u_1}{2O_M(K_1)}) - \exp(-\frac{u_1}{2O_M(K_1)})\right)$$

According to A.1.2,

$$\mathcal{D}(L) \in \mathbf{Z}[\exp(\pm\frac{u_i}{2O_M(K_i)})]_{i=1,\ldots,n}$$

❑

Properties 2.3.2, 2.3.3 and 2.3.4 of the Alexander series are clear from Definition A.1.6 and A.1.3. The symmetry property 2.3.5 of the Alexander series will also be clear from the symmetry property of the Reidemeister torsion

$$\text{(A.1.12)} \quad \tau(N,o(N)) = (-1)^n\iota(\tau(N,o(N)))$$

proved in §A.3.

Definition 2.2.2 of the Alexander series will easily be seen as an application of Definition A.1.6 after §A.2 and §A.3.

§A.4 to §A.6 prove the other properties of the Alexander series listed in §2.3, or their translations in terms of Reidemeister torsion.

A.1.B Outline of the definition of the Reidemeister torsion of link exteriors

NOTATION **A.1.7**: (This notation will be used throughout §A.2 and §A.3.)
N denotes an oriented compact 3-manifold with n tori as boundary ($n \geq 1$).
o(N) denotes an orientation of $H_1(N;\mathbb{R}) \oplus H_2(N;\mathbb{R})$ (as a real vector space).

$\tilde{N}$ denotes the maximal free abelian covering of N, that is the covering corresponding to the natural quotient map:

$$\varphi\colon \pi_1(N) \to Q_{fa}(N) = \frac{H_1(N)}{\text{Torsion}(H_1(N))}$$

$\tilde{*}$ denotes the inverse image of the basepoint $*$ of N under the covering map.

The action of $Q_{fa}(N)$ on $\tilde{N}$ equips $H_1(\tilde{N},\tilde{*};\mathbf{Z})$ with a natural structure of $\mathbf{Z}[Q_{fa}(N)]$-module.

The Reidemeister torsion of N will be defined from presentations of $H_1(\tilde{N},\tilde{*};\mathbf{Z})$ as a $\mathbf{Z}[Q_{fa}(N)]$-module.

DEFINITIONS **A.1.8**: *About module presentations*
Let A be a commutative ring with unit.
A *presentation* of an A-module M is an exact sequence P of A-linear maps of the following type

$$\bigoplus_{i=1}^{r} A\rho_i \xrightarrow{\partial} \bigoplus_{i=1}^{g} A\gamma_i \xrightarrow{\pi} M \to 0$$

A *presentation matrix* of M is a matrix D of an A-linear map ∂ of such a presentation of M. (D has r columns and g rows; the ith column of D contains the coordinates of $\partial(\rho_i)$ with respect to the γ_j's.)
The ρ_i's are the *relators* and the γ_i's are the *generators* associated with P.
The *deficiency* of P is (g-r).
The presentation P is said to be *injective* if ∂ is injective.

Sketch of the definition of the Reidemeister torsion

Step A Exhibit an interesting $\mathbb{Z}[Q_{fa}(N)]$-presentation of deficiency 1 for $H_1(\tilde{N},\tilde{*};\mathbb{Z})$. (See A.2.1.)

Step B Associate with any presentation P of deficiency 1 of $H_1(\tilde{N},\tilde{*};\mathbb{Z})$ a Reidemeister torsion $\tau(N;P)$ defined up to positive units of $\mathbb{Z}[Q_{fa}(N)]$ (see A.2.3 to A.2.8) which satisfies:
(A.2.10)

$$\text{If } Q_{fa}(N) \not\cong \mathbb{Z}, \quad \tau(N;P) \in \mathbb{Z}[\,Q_{fa}(N)\,]$$
$$\text{If } Q_{fa}(N) = \mathbb{Z}g, \quad \tau(N;P)(e^g - 1) \in \mathbb{Z}[\,Q_{fa}(N)\,]$$

and
(A.2.9)

$$\text{If P is non-injective, then } \tau(N;P) = 0.$$

REMARK **A.1.9**: Note that a presentation of deficiency 1 of $H_1(\tilde{N},\tilde{*};\mathbb{Z})$ is non-injective if and only if the rank of $H_1(\tilde{N},\tilde{*};\mathbb{Z})$ is greater than 1. (The rank of $H_1(\tilde{N},\tilde{*};\mathbb{Z})$ is the dimension of the tensor product of $H_1(\tilde{N},\tilde{*};\mathbb{Z})$ by the field of fractions of $\mathbb{Z}[Q_{fa}(N)]$.) So, if some presentation of deficiency 1 of $H_1(\tilde{N},\tilde{*};\mathbb{Z})$ is non-injective, all of them are non-injective and so they define a zero Reidemeister torsion, according to A.2.9.

Step C Define the notion of compatibility of orientation between an injective presentation P of deficiency 1 of $H_1(\tilde{N},\tilde{*};\mathbb{Z})$ and the orientation o(N) of $(H_1(N;\mathbb{R})\oplus H_2(N;\mathbb{R}))$ (see A.2.11 to A.2.16) so that the orientation of an injective presentation of deficiency 1 of $H_1(\tilde{N},\tilde{*};\mathbb{Z})$ is compatible either with o(N) or with -o(N).

DEFINITION **A.1.10**: After making precise the definitions of Step C, we will let the *well-oriented presentations* (of $H_1(\tilde{N},\tilde{*};\mathbb{Z})$) be the injective presentations of deficiency 1 of $H_1(\tilde{N},\tilde{*};\mathbb{Z})$ with orientations compatible with o(N), when o(N) has been given.

Step D Show that two well-oriented presentations of $H_1(\tilde{N},\tilde{*};\mathbb{Z})$ define the same Reidemeister torsion up to positive units (and that an injective presentation of deficiency 1 of $H_1(\tilde{N},\tilde{*};\mathbb{Z})$ having an orientation non-compatible with o(N) defines the opposite one) (see A.2.17 to A.2.24).

Step E (Definition A.2.25) Define the "Reidemeister torsion Up to Positive units" $\tau_{up}(N,o(N))$ of (N,o(N)) as the Reidemeister torsion associated with a well-oriented presentation of $H_1(\tilde{N},\tilde{*};\mathbb{Z})$ if such a presentation exists, and otherwise as zero.

Steps A to D show that $\tau_{up}(N,o(N))$ is well-defined up to multiplication by positive units in $\mathbb{Z}[Q_{fa}(N)]$.
Now, the symmetry property satisfied by this "Reidemeister torsion up to positive units"

$$\text{(A.3.1)} \qquad \tau_{up}(N,o(N)) = (-1)^n\iota(\tau_{up}(N,o(N)))$$

shown in §A.3 allows us to set:

DEFINITION **A.1.11**: *(Reidemeister torsion of a link exterior)*
$\tau(N,o(N))$ is the unique element of the field of fractions of $\mathbb{Z}[\frac{1}{2}Q_{fa}(N)]$ such that $\tau(N,o(N))$ is equal to $\tau_{up}(N,o(N))$ (defined in A.2.25) up to positive units of $\mathbb{Z}[\frac{1}{2}Q_{fa}(N)]$, and $\tau(N,o(N))$ satisfies:

$$\textbf{A.1.12} \qquad \tau(N,o(N)) = (-1)^n\iota(\tau(N,o(N)))$$

So, after §A.3, $\tau(N,o(N))$ will be completely well-defined by A.1.11 and it will be shown to satisfy A.1.2 (from A.2.10) and A.1.3 (from A.2.26).

REMARK **A.1.13**:
The Reidemeister torsion is called Reidemeister torsion even if it is not defined following the standard process to define a Reidemeister torsion described in [Tu]. The definition described in §A.2 is closer to the [C-F] approach to defining Alexander polynomials. (The non-normalized Reidemeister torsion

could be defined as an invariant of the first ideal (see [C-F] Chapter VII Section 4) of the $\mathbb{Z}[Q_{fa}]$-module $H_1(\tilde{N},\tilde{*};\mathbb{Z})$.)

§A.2 Complete definition of the Reidemeister torsion of (N,o(N)) up to positive units

A.2.A Introducing a presentation of $H_1(\tilde{N},\tilde{*};\mathbb{Z})$ over $\mathbb{Z}[Q_{fa}(N)]$ of deficiency 1

A.2.1 Classical arguments of Morse theory show that N strong deformation retracts onto a CW-complex R(N) with the basepoint $*$ of N as unique zero-cell, g one-cells $\gamma_1, ..., \gamma_g$, r two-cells $\rho_1, ..., \rho_r$, and no other cells. Since N has Euler characteristic zero, g must equal (r+1).

The inverse image $R(\tilde{N})$ of R(N), under the covering map from $\tilde{N}$ to N, is thus a CW-complex equivariant under the action of $Q_{fa}(N)$ and is a strong deformation retract of $\tilde{N}$. The only nonzero modules of the chain complex

$$C_*((R(\tilde{N}),\tilde{*});\mathbb{Z}) = C_*(R(N),*;\mathbb{Z}[Q_{fa}(N)])$$

associated with $R(\tilde{N})$ are thus:

- the module $C_1 = C_1(R(N),*;\mathbb{Z}[Q_{fa}(N)])$ free over $\mathbb{Z}[Q_{fa}(N)]$ with basis $\gamma_1, ..., \gamma_{r+1}$, and
- the module $C_2 = C_2(R(N),*;\mathbb{Z}[Q_{fa}(N)])$ free over $\mathbb{Z}[Q_{fa}(N)]$ with basis $\rho_1, ..., \rho_r$.

(See Remark 2.2.3.)

So, $H_1(\tilde{N},\tilde{*};\mathbb{Z})$ is the cokernel of the $\mathbb{Z}[Q_{fa}(N)]$-linear boundary map ∂_2 from C_2 to C_1.

The sequence

$$C_2 \xrightarrow{\partial_2} C_1 \xrightarrow{\pi} H_1(\tilde{N},\tilde{*};\mathbb{Z}) \to 0$$

is thus a presentation of deficiency 1 of $H_1(\tilde{N},\tilde{*};\mathbb{Z})$, and the matrix D of ∂_2 is a presentation matrix of $H_1(\tilde{N},\tilde{*};\mathbb{Z})$.

REMARK **A.2.2**: If a $\mathbb{Z}$-module A (in practice, A = $\mathbb{Z}$, $\mathbb{Q}$ or $\mathbb{R}$) is given the $\mathbb{Z}[Q_{fa}(N)]$-module structure induced by the augmentation morphism ε ($x.a = \varepsilon(x)a$ for $x \in \mathbb{Z}[Q_{fa}(N)]$),

$$(\Rightarrow \mathbb{Z}[Q_{fa}(N)] \otimes_{\mathbb{Z}[Q_{fa}(N)]} A \cong A)$$

then the homology of the tensor product

$$C_*(R(N),*;A) = C_*(R(N),*;\mathbf{Z}[Q_{fa}(N)]) \otimes_{\mathbf{Z}[Q_{fa}(N)]} A$$

is $H_*(N,*;A)$. In particular, $\varepsilon(D)$ is a presentation matrix of $H_1(N,*;A)$ for $A = \mathbf{Z}$, $\mathbb{Q}$ or $\mathbb{R}$.

A.2.B Associating a torsion $\tau(N;P)$ to a presentation P of deficiency 1 of $H_1(\tilde{N},\tilde{*};\mathbf{Z})$

A.2.3 Let P:

$$C_2 = \bigoplus_{i=1}^{r} \mathbf{Z}[Q_{fa}(N)]\rho_i \xrightarrow{\partial_2} C_1 = \bigoplus_{i=1}^{r+1} \mathbf{Z}[Q_{fa}(N)]\gamma_i \xrightarrow{\pi} H_1(\tilde{N},\tilde{*};\mathbf{Z}) \to 0$$

be any presentation of deficiency 1 of $H_1(\tilde{N},\tilde{*};\mathbf{Z})$, and let D be the matrix of ∂_2.

A.2.4 Let ∂_1 denote the composition:

$$C_1 \xrightarrow{\pi} H_1(\tilde{N},\tilde{*};\mathbf{Z}) \xrightarrow{\partial} H_0(\tilde{*};\mathbf{Z}) \xrightarrow{\cong} \mathbf{Z}[Q_{fa}(N)]$$

The map ∂_1 is natural and well-determined as soon as a preferred lifting of $*$ is chosen to determine a preferred isomorphism from $H_0(\tilde{*};\mathbf{Z})$ to $\mathbf{Z}[Q_{fa}(N)]$. We select a lifting for $*$. (Changing the lifting of $*$ multiplies ∂_1 by a positive unit.)

The identity "$\partial_1 \circ \partial_2 = 0$" leads to the relation A.2.5 between the rows $L(\gamma_i)$ ($i=1, \ldots, r+1$) of D:

$$\textbf{A.2.5} \qquad \sum_{i=1}^{r+1} \partial_1(\gamma_i)L(\gamma_i) = 0$$

Let d_i denote the determinant of the matrix obtained from D by deleting the row $L(\gamma_i)$.

From A.2.5, we get:

A.2.6 $\quad (-1)^i\partial_1(\gamma_j)d_i = (-1)^j\partial_1(\gamma_i)d_j$, for any $i,j \in \{1, \ldots, r+1\}$.

Choose $i \in \{1, \ldots, r+1\}$ such that $\partial_1(\gamma_i) \neq 0$, and set

$$\textbf{A.2.7} \qquad \tau(N;P) = \frac{(-1)^{i+r+1}d_i}{\partial_1(\gamma_i)}$$

Relation A.2.6 shows that $\tau(N;P)$ does not depend on the chosen i.

CONCLUSION **A.2.8**

We have thus associated with any presentation P of deficiency 1 of $H_1(\tilde{N},\tilde{*};\mathbf{Z})$ a Reidemeister torsion $\tau(N;P)$ well-defined up to a positive unit of $\mathbf{Z}[Q_{fa}(N)]$.

(The positive unit depends on the choice of the lifting of the basepoint.)

❑

It is clear that:

A.2.9 If P is not injective, $\tau(N;P)$ is zero.

❑

Since ∂_1 is surjective onto the augmentation ideal (kernel of the augmentation morphism) of $\mathbb{Z}[Q_{fa}(N)]$, A.2.6 implies furthermore:

A.2.10

$$\text{If } Q_{fa}(N) \not\cong \mathbb{Z}, (\Rightarrow \operatorname{rank}(Q_{fa}(N)) \geq 2) \quad \tau(N;P) \in \mathbb{Z}[\, Q_{fa}(N)\,]$$

$$\text{If } Q_{fa}(N) = \mathbb{Z}g, \quad \tau(N;P)(e^g - 1) \in \mathbb{Z}[\, Q_{fa}(N)\,]$$

❑

A.2.C Defining the well-oriented presentations (with respect to o(N))

DEFINITION **A.2.11**: *Orientation of a presentation of* $H_1(\tilde{N},\tilde{*};\mathbb{Z})$

An orientation of a free $\mathbb{Z}[Q_{fa}(N)]$-module is an orientation of its tensor product with $\mathbb{R}$ over $\mathbb{Z}[Q_{fa}(N)]$, that is, a basis up to positive transformation (i.e., transformation with a positive unit as determinant). The *orientation of a presentation* P as in A.2.3 is the orientation of the direct sum $C_2 \oplus C_1$ induced by the ordered bases of C_2 and C_1.

The notion of compatibility between the orientation o(N) of $(H_1(N;\mathbb{R}) \oplus H_2(N;\mathbb{R}))$ and the orientation of an injective presentation of deficiency 1 of $H_1(\tilde{N},\tilde{*};\mathbb{Z})$ comes from the following fact:

FACT **A.2.12**

If $\partial_{\mathbb{R}}$ denotes an $\mathbb{R}$-linear map from a finite-dimensional real vector space $C_{\mathbb{R}2}$ to another one $C_{\mathbb{R}1}$, then an orientation of $\operatorname{Coker}(\partial_{\mathbb{R}}) \oplus \operatorname{Ker}(\partial_{\mathbb{R}})$ determines the following orientation of $C_{\mathbb{R}2} \oplus C_{\mathbb{R}1}$:

Orient $C_{\mathbb{R}2}/\operatorname{Ker}(\partial_{\mathbb{R}})$, $\operatorname{Ker}(\partial_{\mathbb{R}})$, $\operatorname{Coker}(\partial_{\mathbb{R}})$, $\operatorname{Im}(\partial_{\mathbb{R}})$, so that the orientation of $\operatorname{Coker}(\partial_{\mathbb{R}}) \oplus \operatorname{Ker}(\partial_{\mathbb{R}})$ is induced by the orientations of $\operatorname{Ker}(\partial_{\mathbb{R}})$ and $\operatorname{Coker}(\partial_{\mathbb{R}})$, and $\partial_{\mathbb{R}}$ induces an oriented isomorphism from $C_{\mathbb{R}2}/\operatorname{Ker}(\partial_{\mathbb{R}})$ onto $\operatorname{Im}(\partial_{\mathbb{R}})$.

Then, orient $C_{\mathbb{R}2} \oplus C_{\mathbb{R}1}$ as the "direct sum"

$$\frac{C_{\mathbb{R}2}}{\operatorname{Ker}(\partial_{\mathbb{R}})} \oplus \operatorname{Ker}(\partial_{\mathbb{R}}) \oplus \operatorname{Im}(\partial_{\mathbb{R}}) \oplus \operatorname{Coker}(\partial_{\mathbb{R}})$$

(This makes sense up to oriented isomorphisms.)

The obtained orientation of $C_{\mathbb{R}2} \oplus C_{\mathbb{R}1}$ is well-defined provided that the orientation of $\mathrm{Coker}(\partial_{\mathbb{R}}) \oplus \mathrm{Ker}(\partial_{\mathbb{R}})$ is well-defined.

❑

Let us admit the following claim, which is clear for the presentation of A.2.1 (see A.2.2).

CLAIM **A.2.13**

Let P be a presentation of $H_1(\tilde{N},\tilde{*};\mathbb{Z})$ as in A.2.3, and let $\partial_{\mathbb{R}}$ be the linear application

$$\partial_{\mathbb{R}} = \partial_2 \otimes_{\mathbb{Z}[Q_{fa}(N)]} \mathrm{Id}_{\mathbb{R}}: \qquad C_2 \otimes_{\mathbb{Z}[Q_{fa}(N)]} \mathbb{R} \quad \rightarrow \quad C_1 \otimes_{\mathbb{Z}[Q_{fa}(N)]} \mathbb{R}$$

There is a canonical isomorphism α_1 *from* $\mathrm{Coker}(\partial_{\mathbb{R}})$ *to* $H_1(N;\mathbb{R})$ *and, if* P *is injective, there is a canonical isomorphism* α_2 *from* $\mathrm{Ker}(\partial_{\mathbb{R}})$ *to* $H_2(N;\mathbb{R})$.

The morphisms α_1 and α_2 are defined in A.2.15 and A.2.16 respectively; they are proved to be isomorphisms in A.2.22.

DEFINITION **A.2.14** *(Compatibility of the orientation of an injective presentation* P *with* o(N) *)*

Keep the notation used in A.2.13, assume that P is injective.

The orientation o(N) induces the orientation carried by $\alpha_1^{-1} \oplus \alpha_2^{-1}$ on $\mathrm{Coker}(\partial_{\mathbb{R}}) \oplus \mathrm{Ker}(\partial_{\mathbb{R}})$. With this orientation we associate the orientation of $C_2 \otimes_{\mathbb{Z}[Q_{fa}(N)]} \mathbb{R} \oplus C_1 \otimes_{\mathbb{Z}[Q_{fa}(N)]} \mathbb{R}$, defined as in A.2.12. If this orientation is the same as the orientation of P, we say that o(N) is *compatible* with the orientation of P; otherwise, we say that o(N) is *non-compatible* with the orientation of P.

A.2.15 DEFINITION OF α_1

α_1 is the surjective morphism induced on $\mathrm{Coker}(\partial_{\mathbb{R}})$ by the composition:

$$C_1 \otimes_{\mathbb{Z}[Q_{fa}(N)]} \mathbb{R} \xrightarrow{\pi \otimes_{\mathbb{Z}[Q_{fa}(N)]} \mathrm{Id}_{\mathbb{R}}} H_1(\tilde{N},\tilde{*};\mathbb{Z}) \otimes_{\mathbb{Z}[Q_{fa}(N)]} \mathbb{R} \xrightarrow{p} H_1(N,*;\mathbb{R})$$

where p is induced by the covering map. ($H_1(N,*;\mathbb{R}) = H_1(N;\mathbb{R})$)

A.2.16 DEFINITION OF α_2 (if P is injective)

Assume that P is injective. Then according to A.1.9, every presentation of $H_1(\tilde{N},\tilde{*};\mathbb{Z})$ is injective, and, we get from the injectivity of a presentation as in A.2.1:

$$H_2(\tilde{N};\mathbb{Z}) = 0$$

Choose, for i = 1, ..., r+1, a curve c_i in $\tilde{N}$ (with boundary in $\tilde{*}$) representing $\pi(\gamma_i)$ in $H_1(\tilde{N},\tilde{*};\mathbb{Z})$.

With an element ρ of C_2 such that

$$\partial_2(\rho) = \sum_{i=1}^{r+1} a_i\gamma_i,$$

we can associate a surface Σ in $\tilde{N}$ with boundary " $\sum_{i=1}^{r+1} a_i c_i$".

If $(\rho\otimes_{\mathbb{Z}[Q_{fa}(N)]} 1) \in \mathrm{Ker}(\partial_{\mathbb{R}})$, then the covering map sends Σ onto a closed surface $p(\Sigma)$ of N. Since $H_2(\tilde{N};\mathbb{Z})$ is trivial, the homology class of $p(\Sigma)$ in $H_2(N;\mathbb{R})$ does not depend on the choice of Σ. It is also independent of the choices of the curves c_i.

We define $\alpha_2(\rho\otimes_{\mathbb{Z}[Q_{fa}(N)]} 1)$ as the homology class of $p(\Sigma)$. This yields a canonical morphism from $\mathrm{Ker}(\partial_{\mathbb{R}})$ to $H_2(N;\mathbb{R})$. It will be shown to be an isomorphism in A.2.22.

A.2.D Proving that two well-oriented presentations define the same Reidemeister torsion (up to positive units)

DEFINITION **A.2.17**: *(+)-isomorphisms of presentations*

We will say that two presentations

$$P = \left(C_2 \overset{\partial}{\to} C_1 \overset{\pi}{\to} M \to 0\right)$$

$$\text{and } P' = \left(C'_2 \overset{\partial'}{\to} C'_1 \overset{\pi'}{\to} M \to 0\right)$$

of a $\mathbb{Z}[Q_{fa}(N)]$-module M are *isomorphic* if there exist two isomorphisms

$$I_1: C_1 \to C'_1 \text{ and } I_2: C_2 \to C'_2$$

such that

$$I_1 o \partial = \partial' o I_2 \text{ and } \pi' o I_1 = \pi$$

Note the naturality of α_1 and α_2 (see A.2.15 and A.2.16) with respect to the isomorphisms of presentations.

Two such presentations are said to be (+)-*isomorphic* if they have furthermore compatible orientations, that is, if the product of the determinants of I_1 and I_2 is a positive unit of $\mathbb{Z}[Q_{fa}(N)]$.

LEMMA **A.2.18**: *Two (+)-isomorphic presentations of deficiency 1 of* $H_1(\tilde{N},\tilde{*};\mathbf{Z})$ *define the same Reidemeister torsion up to positive units.*

PROOF: Let P and P' be two such presentations, as in Definition A.2.17, with

$$C_1 = \bigoplus_{i=1}^{r+1} \mathbf{Z}[Q_{fa}(N)]\gamma_i \qquad C'_1 = \bigoplus_{i=1}^{r+1} \mathbf{Z}[Q_{fa}(N)]\gamma'_i$$

To prove A.2.18, it suffices to study the case:

I_2 = Identity, ($\det(I_1)$ is thus a positive unit), $\partial_1(\gamma_{r+1}) \neq 0$, $\partial'_1(\gamma'_{r+1}) \neq 0$.

(We use the notation of A.2.B and, D and I_1 denote the matrices of ∂ and I_1. $D' = I_1 D$)

$$\partial'_1(\gamma'_{r+1}) = (\partial_1(\gamma_1), \ldots, \partial_1(\gamma_{r+1}))(\text{column number } (r+1) \text{ of } I_1^{-1})$$

$$= \sum_{i=1}^{r+1} \partial_1(\gamma_i)\left(I_1^{-1}\right)_{i,r+1}$$

According to A.2.6,

$$\partial'_1(\gamma'_{r+1})d_{r+1} \quad = \partial_1(\gamma_{r+1}) \sum_{i=1}^{r+1} (-1)^{i+1+r} d_i \left(I_1^{-1}\right)_{i,r+1} = \partial_1(\gamma_{r+1}) \det\left(D, I_1^{-1}\begin{pmatrix}0\\ \cdots\\ 0\\ 1\end{pmatrix}\right)$$

$$= \partial_1(\gamma_{r+1}) \det(I_1)^{-1} \det\left(I_1 D, \begin{pmatrix}0\\ \cdots\\ 0\\ 1\end{pmatrix}\right) = \partial_1(\gamma_{r+1}) \det(I_1)^{-1} d'_{r+1}$$

This proves A.2.18.

❑

DEFINITION **A.2.19**: *Oriented stabilization of a presentation of* $H_1(\tilde{N},\tilde{*};\mathbf{Z})$

Consider the presentation P of $H_1(\tilde{N},\tilde{*};\mathbf{Z})$:

$$C_2 = \bigoplus_{i=1}^{r} \mathbf{Z}[Q_{fa}(N)]\rho_i \xrightarrow{\partial_2} C_1 = \bigoplus_{i=1}^{r+1} \mathbf{Z}[Q_{fa}(N)]\gamma_i \xrightarrow{\pi} H_1(\tilde{N},\tilde{*};\mathbf{Z}) \to 0$$

Let $\tilde{\pi}(\gamma_0)$ be an element of $H_1(\tilde{N},\tilde{*};\mathbf{Z})$ which can be written as

$$\tilde{\pi}(\gamma_0) = \sum_{i=1}^{r+1} a_i \pi(\gamma_i)$$

The following presentation $\tilde{P}$ of $H_1(\tilde{N},\tilde{*};\mathbf{Z})$:

$$\tilde{C}_2 = \mathbf{Z}[Q_{fa}(N)]\rho_0 \oplus C_2 \xrightarrow{\tilde{\partial}_2} \tilde{C}_1 = \mathbf{Z}[Q_{fa}(N)]\gamma_0 \oplus C_1 \xrightarrow{\tilde{\pi}} H_1(\tilde{N},\tilde{*};\mathbf{Z}) \to 0$$

where

$$\tilde{\partial}_2(\rho_0) = \gamma_0 - \sum_{i=1}^{r+1} a_i \gamma_i$$

and $\tilde{\pi}$ and $\tilde{\partial}_2$ extend π and ∂_2, is an *oriented stabilization* of P. An *oriented stabilization* of P is a presentation of $H_1(\tilde{N},\tilde{*};\mathbf{Z})$ obtained from P by iterating this kind of modification.

The following lemma is immediate:

LEMMA **A.2.20**: *An oriented stabilization does not change the Reidemeister torsion, and an oriented stabilization of a well-oriented presentation is well-oriented.*

❑

LEMMA **A.2.21**: *Two injective presentations of deficiency 1 of* $H_1(\tilde{N},\tilde{*};\mathbf{Z})$ *have isomorphic oriented stabilizations.*

PROOF: Let P and P' be two injective presentations of deficiency 1 of $H_1(\tilde{N},\tilde{*};\mathbf{Z})$.

(We use the notation of A.2.3 for P and similar notation for other presentations.) Use oriented stabilizations to transform P into a presentation $\tilde{P}$ with module of generators

$$\tilde{C}_1 = C'_1 \oplus C_1$$

and P' into a presentation $\tilde{P}'$ with

$$\tilde{C}'_1 = C_1 \oplus C'_1$$

Let I_1 be the natural isomorphism (permuting the generators) from $\tilde{C}_1$ to $\tilde{C}'_1$. The maps $I_1 \circ \tilde{\partial}_2$ and $\tilde{\partial}'_2$ have the same image $\mathrm{Ker}(\pi \oplus \pi')$ in $\tilde{C}'_1$ and they are injective because the presentations are injective. This induces a natural isomorphism between the modules of relators of $\tilde{P}$ and $\tilde{P}'$ and shows that $\tilde{P}$ and $\tilde{P}'$ are isomorphic.

❑

PARENTHESIS **A.2.22**: A.2.21 proves in particular that α_1 and α_2 are isomorphisms if P is injective (since this property is true for a presentation as in A.2.1 and is preserved under isomorphisms and oriented stabilizations).

LEMMA **A.2.23**: *Two well-oriented presentations of* $H_1(\tilde{N},\tilde{*};\mathbf{Z})$ *have (+)-isomorphic oriented stabilizations.*

Indeed, the isomorphism between their two oriented stabilizations defined in the proof of A.2.21 must preserve orientation.

❑

Lemmas A.2.18 , A.2.20 and A.2.23 give the conclusion of Step D:

CONCLUSION **A.2.24**: *Two well-oriented* (with respect to o(N)) *presentations of* $H_1(\tilde{N},\tilde{*};\mathbb{Z})$ *define the same Reidemeister torsion (up to positive units).*

❑

A.2.E Defining the Reidemeister torsion up to positive units

DEFINITION **A.2.25**

Now, (according to A.1.9 and A.2.24), it makes sense to define $\tau_{up}(N, o(N))$ up to positive units as the Reidemeister torsion $\tau(N;P)$ defined (by A.2.7) from a well-oriented (with respect to o(N)) presentation P of $H_1(\tilde{N},\tilde{*};\mathbb{Z})$ if such a presentation exists, and as zero otherwise.

Furthermore, it is clear that, up to positive units

A.2.26 $\tau_{up}(N, -o(N)) = -\tau_{up}(N, o(N))$

❑

EXAMPLE **A.2.27**:

Let $L = K_0, K_1, \ldots, K_n$ denote the following link in S^3:

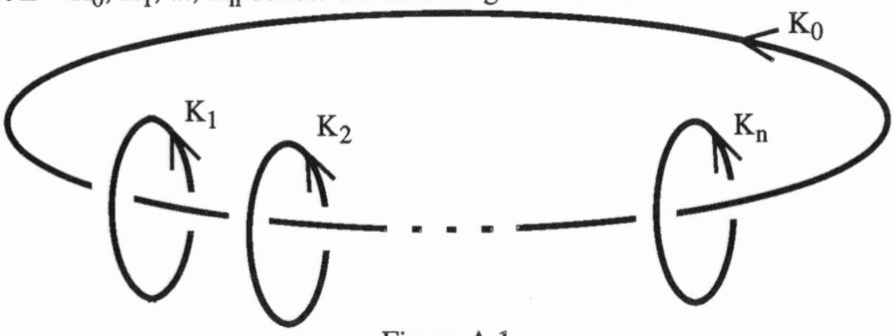

Figure A.1

Its complement strong deformation retracts onto the product of the wedge of the meridians $(m_i)_{i=1,\ldots,n}$ of the components K_i by a circle which is the meridian m_0 of K_0. Give this product the following cellular decomposition:

- the basepoint ● as unique zero-cell,
- the meridians $m_0, m_1, m_2, \ldots, m_n$ as one-cells and,
- n two-cells T_i, $i = 1, \ldots, n$, representing the oriented boundaries of the tubular neighborhoods of the K_i.

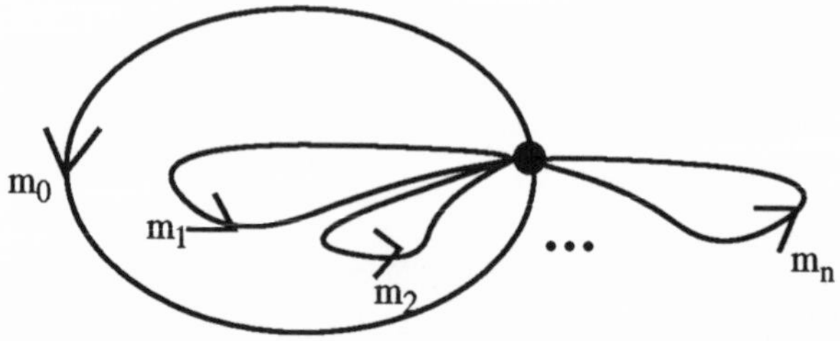

Figure A.2: One-skeleton of the given strong deformation retract of $S^3 \backslash L$

With the choices of the liftings indicated by Figure A.3 below, the boundary of T_i in the associated cellular chain complex with coefficients in $\mathbf{Z}[Q_{fa}(S^3 \backslash L)]$ is:

$$(\exp(m_i) - 1)m_0 + (1 - \exp(m_0))m_i$$

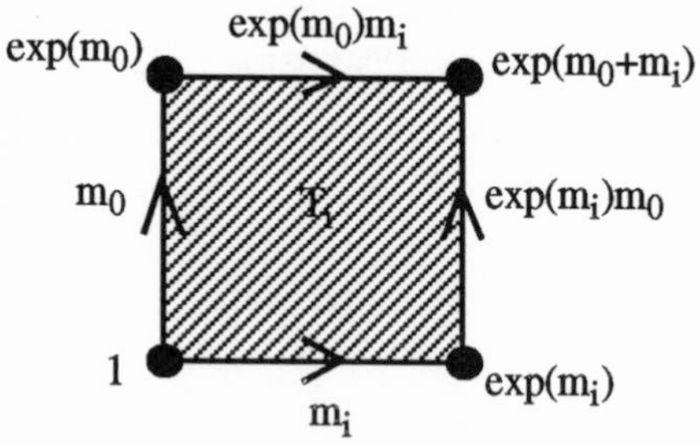

Figure A.3: A lifting of T_i in the maximal free abelian covering of $S^3 \backslash L$

The matrix of the second boundary map in this cellular complex with respect to the bases $(T_1, ..., T_n)$ and $(m_1, ..., m_n, m_0)$ is then:

$$\begin{pmatrix} 1 - \exp(m_0) & 0 & \cdots & 0 \\ 0 & 1 - \exp(m_0) & \cdots & 0 \\ \cdots & \cdots & \cdots & \cdots \\ 0 & 0 & \cdots & 1 - \exp(m_0) \\ \exp(m_1) - 1 & \exp(m_2) - 1 & \cdots & \exp(m_n) - 1 \end{pmatrix}$$

The considered presentation is well-oriented with respect to $o_L(S^3 \backslash L)$ (see A.1.5). So,

$$\tau_{up}(S^3 \backslash L, o_L(S^3 \backslash L)) = (-1)^n \left(\exp(m_0) - 1\right)^{n-1}$$

and, according to Definition A.1.11,

$$\tau(S^3 \backslash L, o_L(S^3 \backslash L)) = (-1)^n \left(\exp(\frac{1}{2} m_0) - \exp(- \frac{1}{2} m_0)\right)^{n-1}$$

❑

§A.3 Proof of the symmetry property of the Reidemeister torsion

We now prove the (well-known):

SYMMETRY PROPERTY **A.3.1**
Let N *be the exterior of an n-component link in a closed orientable 3-manifold and let* o(N) *be an orientation of* $H_1(N;\mathbb{R})\oplus H_2(N;\mathbb{R})$, *then*

$$\tau_{up}(N,o(N))=(-1)^{n}\iota(\tau_{up}(N,o(N)))$$

(up to positive units of $\mathbb{Z}[Q_{fa}(N)]$)

The proof consists in finding two "dual" cellular decompositions for N (paying attention to the difference between duality on the boundary and duality in the interior of N), so that computing τ_{up} with one presentation or the other gives conjugate results.

A.3.A A symmetric presentation of N

Write N as a closed 3-manifold M minus an open tubular link neighborhood $\overset{o}{T}(L)$, where:

$$L = \{K_1, \ldots, K_n\}$$

Choose a Heegaard splitting of M:

$$M = P\cup_{h(\partial P)=\partial\Pi}\Pi$$

such that the link L lies on ∂P.
(P and Π are two handlebodies, h is the gluing homeomorphism from ∂P to $\partial\Pi$.)
(To obtain such a decomposition, start with a tubular neighborhood of L, join its different components to get a handlebody, add one-handles to this handlebody so that its complement also becomes a handlebody, and push L onto the boundary.)

Call $\overset{o}{N}_{\partial P}(L)$ an annular neighborhood of L in ∂P, and call $\overset{o}{N}_{\partial\Pi}(L)$ its image under h. Then,

$$N = P \underset{h(\,\partial P\setminus\overset{o}{N}_{\partial P}(L)\,) = \partial\Pi\setminus\overset{o}{N}_{\partial\Pi}(L)}{\cup} \Pi$$

This is the symmetric presentation for M, which will show the symmetry of the Reidemeister torsion.

A.3.B Describing two "dual" cellular decompositions for N

Write each K_i as the union of two arcs c_{iP} and γ_{iP} lying on ∂P, and write a neighborhood of K_i in ∂P as the union of two rectangular neighborhoods $N(c_{iP})$ and $N(\gamma_{iP})$ of c_{iP} and γ_{iP} as indicated by the following picture in ∂P:

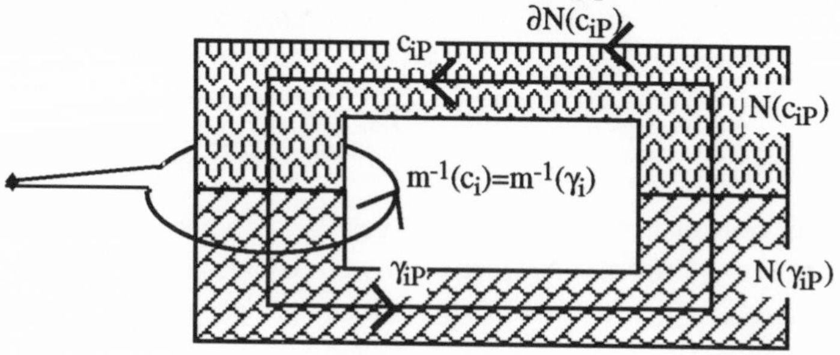

Figure A.4

Set $c_{i\Pi} = h(c_{iP})$, $\gamma_{i\Pi} = h(\gamma_{iP})$, $N(c_{i\Pi}) = h(N(c_{iP}))$, and $N(\gamma_{i\Pi}) = h(N(\gamma_{iP}))$.
Let $\blacklozenge \in P\backslash N_{\partial P}(L)$ be the basepoint of N.
Let g be the genus of ∂P.

Embed g oriented disks D_i, i=1, ..., g, in the handlebody P so that:

- $B(P) = P\backslash \bigcup_{i=1}^{g} D_i$ is a ball (minus 2g disks on the boundary), and,
- the oriented boundary ∂D_i of D_i lies on $\left(\partial P \setminus \left(\blacklozenge \cup \left(\cup_{i=1}^{n} N(c_{iP}) \right)\right)\right)$

  and intersects $N(\gamma_{jP})$ along segments "orthogonal" to γ_{jP}, for all j.

Let a_i be "the" loop in P, based on $\blacklozenge$, intersecting D_i once transversally with a sign +, and otherwise lying in the interior of B(P).

Define similarly g disks Δ_i in Π, the $\partial\Delta_i$'s lying on $\partial\Pi$, the ball $B(\Pi)$, and the α_i's.

$$\partial\Delta_i \subset \partial\Pi \setminus (\blacklozenge \cup (\cup_{i=1}^{n} N(\gamma_{i\Pi})))$$

For $i \in \{1, \ldots, n\}$, let $m^{-1}(c_i)=m^{-1}(\gamma_i)=a_{i+g}=\alpha_{i+g}$ be a negatively oriented meridian of c_i (and γ_i) based on $\blacklozenge$ as shown in Figure A.4, and let Δ_{i+g} (respectively D_{i+g}) be the disk $N(\gamma_{i\Pi})$ oriented as part of $\partial\Pi$ (respectively the disk $N(c_{iP})$ oriented as part of ∂P) .

$P \cup (\cup_{i=1}^{n} N(c_{i\Pi}))$ strong deformation retracts onto the wedge of the $(a_i)_{i=1,\dots,n+g}$.
This wedge will be the one-skeleton C^1 of our first cellular decomposition C.
Gluing the $(\Delta_i)_{i=1,\dots,n+g-1}$ on $P \cup (\cup_{i=1}^{n} N(c_{i\Pi}))$ yields a strong deformation retract of N. These disks will be the two-cells of the decomposition C.
(REMARK: To get a genuine CW-complex R(C), it would now suffice to perform the strong deformation retraction of $P \cup (\cup_{i=1}^{n} N(c_{i\Pi}))$ onto C^1; but we prefer to consider that the two-cells are attached to $P \cup (\cup_{i=1}^{n} N(c_{i\Pi}))$, and we naturally associate with this "cellular decomposition" C, the cellular chain complex associated with R(C).)

Define the "dual" decomposition Γ in the same way. The one-cells of Γ are the $(\alpha_i)_{i=1,\dots,n+g}$, its two-cells are the $(D_i)_{i=1,\dots,n+g-1}$.

A.3.C Stating the relation between the presentation matrices associated with the two dual cellular decompositions of N

A.3.2 Assume, without loss of generality, that the vector space freely generated by the meridians of L is injected into $H_1(N;\mathbb{R})$.

Denote by $\frac{\partial \Delta_j}{\partial a_i}$ the ith coordinate of the boundary $\partial\Delta_j$ of a two-cell Δ_j with respect to the basis of one-cells $(a_i)_{i=1,\dots,n+g}$ in the equivariant chain complex associated with C.

Let $F = [f_{ij}]_{1\le i,j\le n+g}$, and $\Phi=[\phi_{ij}]_{1\le i,j\le n+g}$ be the matrices defined by:

- If $i \le g$, $\quad f_{ij} = \frac{\partial \Delta_j}{\partial a_i} \qquad \phi_{ij} = \frac{\partial D_j}{\partial \alpha_i}$
- If $i = g+k$, $k > 0$, $\quad f_{ij} = \frac{\partial \Delta_j}{\partial a_i}(1-\exp[m^{-1}(K_k)]) \qquad \phi_{ij} = \frac{\partial D_j}{\partial \alpha_i}(1-\exp[m^{-1}(K_k)])$.

Let $d_{n+g}(F)$ (respectively $d_{n+g}(\Phi)$) be the cofactor of $f_{(n+g)(n+g)}$ in F (respectively of $\phi_{(n+g)(n+g)}$ in Φ). Then, according to A.2.8, the following equalities hold, up to positive units in $\mathbf{Z}[Q_{fa}(N)]$:

$$\textbf{A.3.3} \qquad \tau(N;C) = \frac{d_{n+g}(F)}{\prod_{i=1}^{n}(1-\exp[m^{-1}(K_i)])} \quad \text{and} \quad \tau(N;\Gamma) = \frac{d_{n+g}(\Phi)}{\prod_{i=1}^{n}(1-\exp[m^{-1}(K_i)])}$$

We will prove in Subsection A.3.F that, on the more precise assumptions regarding the cellular decompositions of Subsection A.3.D, we have:

$$\textbf{A.3.4} \qquad F = \iota({}^t\Phi)$$

where ${}^t\Phi$ is the transposed matrix of Φ.
A.3.4 shows in particular that

A.3.5 C and Γ have compatible orientations.

(PROOF OF A.3.5:
If $\varepsilon(f_{ij}) = 0 = \varepsilon(\phi_{ij})$, for any (i,j) such that i or j is greater than $(k = g - \beta_1(M))$, according to A.3.4,

$$\det[f_{ij}]_{1\le i,j\le k} = \det[\phi_{ij}]_{1\le i,j\le k}\ (\neq 0)$$

and it suffices to compare the orientations of the bases ($a_* = (a_{k+1},\ldots,a_g)$, $\Delta_* = (\Delta_{k+1},\ldots,\Delta_g)$) and ($\alpha_* = (\alpha_{k+1},\ldots,\alpha_g)$, $D_* = (D_{k+1},\ldots,D_g)$) of $H_1(M;\mathbb{R})\oplus H_2(M;\mathbb{R})$.
Since a_* and α_* are respectively dual to D_* and Δ_* for the intersection in M by construction, the change of basis from a_* to α_* and the change of basis from Δ_* and D_* have the same sign. This proves A.3.5 in this particular case, to which any case is reduced by means of (+)-isomorphisms (see A.2.17). ❑)
So, we can assume that C and Γ both have an orientation compatible with o(N).
Now, the symmetry property A.3.1 follows from A.3.3 and A.3.4.

A.3.D Specifying the dual decompositions

Consider the decomposition C.
Assume that the glued Δ_i, for i = 1, ..., g, is the initial one pushed slightly in the direction of its negative normal. Similarly, assume that the boundary of $N(\gamma_{i\Pi}) = \Delta_{i+g}$ is pushed slightly out of $N(\gamma_{i\Pi})$ on $P\setminus N(\gamma_{iP})$. That is, the boundaries of the two-cells Δ_i are pushed slightly to their right-hand side on $\partial\Pi$ and to their left-hand side on ∂P. We think locally of an oriented surface as part of our sheet of paper oriented as usual.

Attach the boundaries of the Δ_i's to ♦ with a path in $B(\Pi)$ $(\cap\partial P\setminus N(\gamma_{jP})_{j=1,\ldots,n})$.
Do the same for Γ.

A.3.E Computing the presentation matrices on the splitting surface

To compute $\frac{\partial\Delta}{\partial a}$, it suffices to follow the boundary $\partial\Delta$ of Δ on

$$\left(\partial P\setminus N_{\partial P}(L)\right)\cup\left(\bigcup_{i=1}^{n} N(c_{i\Pi})\right)$$

(Here, $N(c_{i\Pi})$ is (exceptionally) oriented as $h(N(c_{iP}))$.)

and to count the contribution of each crossing X of $\partial\Delta$ with a^*, where

$$a_i^* = \partial D_i \text{ if } i \leq g \text{ and } a_{i+g}^* = c_{i\Pi}.$$

(When $\partial\Delta$ does not meet the $(a_i^*)_{i=1,\ldots,n+g}$ it remains in the ball B(P) containing the basepoint. Note also that, if $\partial\Delta$ crossed the same curve twice in opposite directions at the same point, and if the restriction of $\partial\Delta$ betweeen these two crossings were null-homologous in N, then the contributions of the two crossings would cancel out. This allows us to remain relatively vague about the way of choosing paths to join our boundaries to ♦.)

Call a crossing X positive or negative depending on how it looks in the oriented surface ∂P (see Figure A.5):

Positive crossing in ∂P Negative crossing in ∂P

Figure A.5

If C is a point of ∂P outside $N_{\partial P}(L)$, the $\partial(\Delta_i)$'s, and the $\partial(D_i)$'s, let $\beta(C) \in Q_{fa}(N)$ denote the class of the loop in N which goes from ♦ to C by B(Π) and returns to ♦ by B(P), and let

$$b(C) = -\beta(C)$$

Thus, the contribution to $\frac{\partial\Delta}{\partial a}$ of the positive crossing of Figure A.5 is $\exp(\beta(B))$ while the contribution of the negative crossing is $-\exp(\beta(A))$. (As soon as a preferred lifting ♦♦ for the basepoint is chosen, the preferred liftings of the one-cells and the boundaries of the preferred liftings of the two-cells start from ♦♦.)

The computation of $\frac{\partial D_i}{\partial \alpha_j}$ is performed similarly.

A.3.F Proving A.3.4

A.3.6 If $i,j \leq g$,

$$\iota\left(\frac{\partial\Delta_j}{\partial a_i}\right) = \frac{\partial D_i}{\partial \alpha_j}$$

PROOF OF A.3.6: In this case, $\frac{\partial\Delta_j}{\partial a_i}$ and $\frac{\partial D_i}{\partial \alpha_j}$ are the sums of the contributions of the crossings of $\partial\Delta_j$ and ∂D_i. The contribution to $\frac{\partial\Delta_j}{\partial a_i}$ and $\frac{\partial D_i}{\partial \alpha_j}$, of a positive crossing as in Figure A.5, is $\exp(\beta(B))$ and $\exp(b(B))$, respectively. It is $[-\exp(\beta(A))]$ and $[-\exp(b(A))]$ respectively for a negative crossing as in Figure A.5. Thus, they are conjugate under the involution ι in both cases. (Note that, since $\frac{\partial D_i}{\partial \alpha_j}$ is computed with the orientation induced by $\partial\Pi$, the crossing is given the same sign in both computations.)

❑

A.3.7 If $j \le g$, for $1 \le i \le n$,

$$\iota((1-\exp[m^{-1}(K_i)])\frac{\partial \Delta_j}{\partial a_{i+g}}) = \frac{\partial D_{i+g}}{\partial \alpha_j}$$

PROOF OF A.3.7:

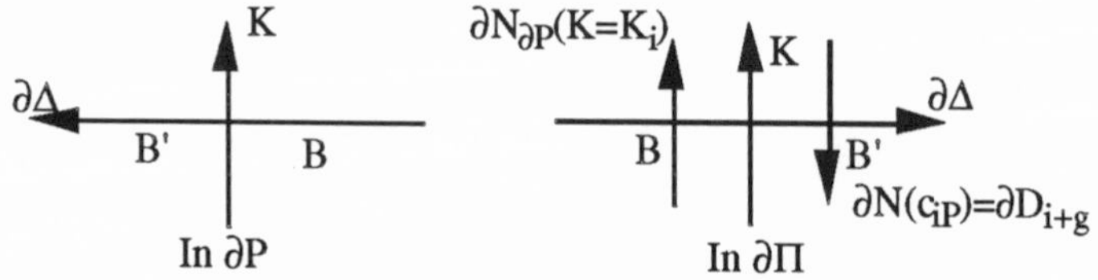

Figure A.6: First kind of crossing

The first kind of crossing shown in Figure A.6 contributes

to $\frac{\partial \Delta_j}{\partial a_{i+g}}$ by $\exp(\beta(B))$ and,

to $\frac{\partial D_{i+g}}{\partial \alpha_j}$ by $(\exp(b(B)) - \exp(b(B')))$.

So, since $\beta(B') = m^{-1}(K_i) + \beta(B)$, this crossing contributes in the same way to the two sides of Equality A.3.7.

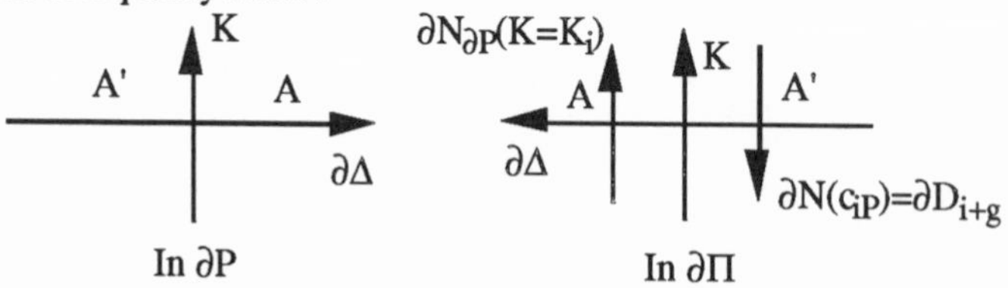

Figure A.7: Second kind of crossing

Similarly, the contribution of the second kind of crossing (shown in Figure A.7) to the two sides of Equality A.3.7 is $(\exp(b(A')) - \exp(b(A)))$. This proves A.3.7.

❑

The symmetric case is symmetric. So, to prove A.3.4, we are left with the proof of A.3.8.

A.3.8 If $1 \leq i, j \leq n$,

$$\iota((1 - \exp[m^{-1}(K_j)]) \frac{\partial \Delta_{j+g}}{\partial a_{i+g}}) = (1 - \exp[m^{-1}(K_j)]) \frac{\partial D_{i+g}}{\partial \alpha_{j+g}}$$

PROOF OF A.3.8: Both sides of Equality A.3.8 are zero when i and j are different. So, we assume i=j and we prove that both sides of Equality A.3.8 equal

$$\exp(b(B)) - \exp(b(B')) + \exp(b(A')) - \exp(b(A))$$

with the points A, A', B and B' of Figure A.8.

The first part of Figure A.8 shows the computation of $\frac{\partial \Delta_{i+g}}{\partial a_{i+g}}$ while the second part shows that the result is symmetric.

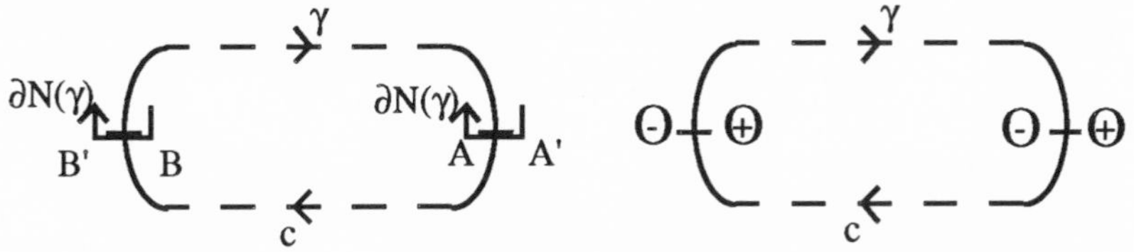

Figure A.8: In ∂P

❑

This proves A.3.4 and completes the proof of the symmetry property A.3.1.

❑ ❑

From now on, the Reidemeister torsion is well-defined by Definition A.1.11.

§A.4 Various properties of the Reidemeister torsion

The properties below correspond to Properties 2.3.6 to 2.3.9 of the Alexander series.

A.4.1 *Restriction formula*

Let N *be the exterior of an n-component link with* $n \geq 2$.

Let T *be an oriented component of the boundary of* (-N). *Let* m *and* ℓ *be oriented primitive curves on* T *such that their intersection number* $\langle m, \ell \rangle$ *equals one.*

Let N_m *be defined by:*

$$N_m = N \cup_h D^2 \times S^1$$

where h *is a homeomorphism from* T *onto* $\partial(D^2 \times S^1)$ *which maps* m *to the meridian of* $D^2 \times S^1$.

Let α_m *be the morphism from* $\mathbb{Z}[Q_{fa}(N)]$ *to* $\mathbb{Z}[Q_{fa}(N_m)]$ *induced by the inclusion.*

Then, if the orientations $o(N)$ *of* $H_1(N;\mathbb{R})\oplus H_2(N;\mathbb{R})$ *and* $o(N_m)$ *of* $H_1(N_m;\mathbb{R})\oplus H_2(N_m;\mathbb{R})$ *are related as indicated below:*

$$\alpha_m(\tau(N, o(N))) = \left(\exp(-\tfrac{1}{2}[\ell]) - \exp(\tfrac{1}{2}[\ell])\right)\tau(N_m, o(N_m))$$

$$([\ell] \in H_1(N_m;\mathbb{Z}))$$

Required relations between $o(N)$ *and* $o(N_m)$ *to get the correct sign in the restriction formula*

Assume that the orientations $o(N)$ of $H_1(N;\mathbb{R})\oplus H_2(N;\mathbb{R})$ and $o(N_m)$ of $H_1(N_m;\mathbb{R})\oplus H_2(N_m;\mathbb{R})$ are respectively induced by orientations of $H_1(N;\mathbb{R})$, $H_2(N;\mathbb{R})$ and of $H_1(N_m;\mathbb{R})$, $H_2(N_m;\mathbb{R})$ and that the following isomorphisms (canonical up to oriented isomorphisms) respect these orientations:

- If m does not represent zero in $H_1(N;\mathbb{R})$
  - $H_1(N;\mathbb{R}) \cong \mathbb{R}[m]\oplus H_1(N_m;\mathbb{R})$
  - $H_2(N;\mathbb{R}) \cong \mathbb{R}[T]\oplus H_2(N_m;\mathbb{R})$
- If m represents zero (this case is not actually needed in the book, here $H_2(N;\mathbb{R})/\mathbb{R}[T]$ is also supposed to be oriented)
  - $H_1(N;\mathbb{R}) \cong H_1(N_m;\mathbb{R})$
  - $H_2(N;\mathbb{R}) \cong \mathbb{R}[T] \oplus \left(H_2(N;\mathbb{R})/\mathbb{R}[T] \right)$
  - $H_2(N_m;\mathbb{R}) \cong \mathbb{R}[S] \oplus \left(H_2(N;\mathbb{R})/\mathbb{R}[T] \right)$, where S is an oriented surface embedded in the interior of N_m such that the intersection number $\langle S,\ell\rangle$ is positive.

(Here the brackets denote homology classes.)

PROOF: Choose a CW-complex, strong deformation retract of N, as in A.2.1, and such that there are two one-cells [m] and [ℓ] representing m and ℓ and one two-cell [T] representing T. ([T] is attached along $m\ell m^{-1}\ell^{-1}$.)
Then, in $C_*(R(N),*;\mathbb{Z}[Q_{fa}(N)])$:

$$\partial_2([T]) = [m] + e^{[m]}[\ell] - e^{[\ell]}[m] - [\ell] = (1 - e^{[\ell]})[m] + (e^{[m]} - 1)[\ell]$$

$$(\text{where } \alpha_m(e^{[m]} - 1)=0)$$

Now, replacing the cell [T] by a disk D(m) with boundary m in the previous CW-complex yields a CW-complex, strong deformation retract of N_m.
Call D and D_m the matrices associated with these decompositions as in A.2.1.
Then $\alpha_m(D) = D_m$ except for the only nonzero term in the column of [T] or [D(m)]. This term is their respective coordinate with respect to [m], which is $(1 - e^{[\ell]})$ or 1.
According to the relations between $o(N)$ and $o(N_m)$, the two presentations are either both well-oriented or both not well-oriented (see A.2.C).
So, up to positive units:

$$\alpha_m(\tau(N,o(N))) = (1-e^{[\ell]})\tau(N_m,o(N_m))$$

This proves A.4.1.

❑

A.4.2 *Nullity on split exteriors*

If there exists a sphere S^2, *embedded in the interior of a link exterior* N, *which separates* N *into two components, both of them intersecting the boundary of* N, *then*

$$\tau(N,o(N)) = 0$$

PROOF: It suffices to see that, in this case, there is a CW-complex, strong deformation retract of N, as in A.2.1 such that one of its two-cells represents the separating sphere. The boundary ∂_2 of such a cell will thus be zero in the associated chain complex.

❑

Property 2.3.8 can be written as follows

Let L be an oriented link in an oriented rational homology sphere M, and let K_0 be an oriented meridian of a component K_i of L.

Removing from M a tubular neighborhood of L containing K_0 induces an embedding of the exterior of L in $M\backslash(L\cup K_0)$. Let

$$\mathcal{E}: \mathbb{Z}[Q_{fa}(M\backslash L)] \rightarrow \mathbb{Z}[Q_{fa}(M\backslash(L\cup K_0)]$$

be the morphism induced by this embedding, then

A.4.3

$$\tau(\,M\backslash(L\cup K_0)\,,\,o_{L\cup K_0}(M\backslash(L\cup K_0))\,)$$

$$= \left(\exp(-\frac{m_i}{2}) - \exp(\frac{m_i}{2})\right)\mathcal{E}(\,\tau(\,M\backslash L\,,\,o_L(M\backslash L)\,)\,)$$

PROOF OF A.4.3 OR 2.3.8: Choose a CW-complex C, strong deformation retract of the exterior of L (as in A.2.1), with a meridian m_i of K_i as a one-cell. Adding a one-cell m_0 to C and gluing a two-cell e_0 along $m_0m_im_0^{-1}m_i^{-1}$ yields a strong deformation retract C' of the exterior of $L\cup K_0$. (The two-cell e_0 represents the tubular neighborhood of K_0.) No two-cell of C' except e_0 has m_0 in its boundary and the coordinate of $\partial_2(e_0)$ with respect to m_0 is $(1 - \exp(m_i))$. Therefore, up to units,

$$\frac{\tau(M\backslash(L\cup K_0))}{\mathcal{E}(\tau(M\backslash L))} = (1 - \exp(m_i))$$

To get the normalization, check the orientation or use the restriction formula.

❑

PROPERTY **2.3.9**

If the components of a link $L = \{K_1, \ldots, K_n\}$ $(n \geq 2)$ *in a rational homology sphere are null-homologous, then*

$$\mathcal{D}(L) \in \prod_{i=1}^{n} \exp\left(\frac{u_i}{2}\left(\sum_{j \in \{1,\ldots,n\}\setminus\{i\}} \ell_{ij} + 1\right)\right) \mathbf{Z}[\exp(\pm u_i)]_{i=1,\ldots,n}$$

A.4.4 PROOF OF 2.3.9: §A.6 will prove in particular that, for any null-homologous knot K,

$$\Delta(K) \in \mathbf{Z}[t^{\pm 1}]$$

Property 2.3.9 is then clear when all the linking numbers in L are nonzero, from the restriction formulae 2.3.6.

Otherwise, we can add a trivial component K_0 to L which links every initial component of L, twist the exterior of K_0 (as in 1.6.4) an [even] number of times to get a link where all linking numbers are nonzero [and unchanged mod 2] (see 3.1.3), and then get the result for the initial link from 2.3.2 and 2.3.6.

❑

§A.5 A systematic way of computing the Alexander polynomials of links in S^3

(or the comparison with the Hartley and the Boyer and Lines normalizations)

A normalization of the Alexander polynomial of oriented links in S^3 as well as an explicit way of computing it is given in [Hart].

§A.5 is devoted to explaining the Hartley normalization of the Alexander polynomial of links in S^3 with the point of view of this book, and to proving the following relation between the two normalizations:

A.5.1 *If* L *is an oriented link of n components in* S^3 *with* $n \geq 2$,

$$\Delta(L) = (-1)^{n-1} \Delta_{\text{Hartley}}(L)$$

The relation between Alexander polynomials of links with homeomorphic exteriors in [B-L 2] and Property 2.3.2 are thus sufficient to prove completely:

2.3.11 *If* L *is an oriented link of n components in a rational homology sphere* M *with* $n \geq 2$,

$$\Delta(L) = |H_1(M;\mathbf{Z})|\,(-1)^{n-1} \Delta_{\text{Boyer-Lines}}(L)$$

❑

A.5.A The Wirtinger cellular decomposition of a link exterior in S^3

With a regular planar projection of a link L in S^3, we can associate the following cellular decomposition of $S^3 \setminus \overset{o}{T}(L)$. (We will only consider the case when each component crosses under. Otherwise the link is either a split link or the trivial knot.)

See S^3 as the Euclidian space $\mathbb{R}^3$ plus a point $*$ and assume that the projection of L is the vertical projection onto the horizontal plane H.

Assume that L is above H except near the crossings of the projection where the under-arc goes under H.

The intersection of L's exterior with the upper half-space above H (plus a neighborhood of *) strong deformation retracts onto the wedge of the meridians of the presentation over-arcs. (The over-arcs are the connected components of L's diagram, this diagram is cut where the link crosses under.) This wedge will be the one-skeleton X^1 of the decomposition.

With a crossing P, associate a disk D(P) such that D(P) projects (one-to-one) onto H as a neighborhood of the part of L which goes under H near P, D(P) is under L and H, and the boundary of D(P) lies on H. See D(P) as a part of the boundary of the tubular neighborhood of L.

Attaching all the two-cells D(P) except one to the part above H of L's exterior, and retracting this upper part on X^1 yields a CW-complex, strong deformation retract of L's exterior, from which we can compute the Alexander polynomial of L as in §A.2.

Orient the one-cells as oriented meridians of L, and the two-cells as part of the boundary of L's exterior (so, the oriented boundary of D(P) seen from above H turns counterclockwise around P).

A.5.B Obtaining the normalization from the Wirtinger decomposition

Hartley obtains the normalization by numbering the cells of the decomposition according to the rule:

Number the k crossings $P_1, \ldots, P_k$ of the projection from 1 to k arbitrarily, and number the over-arcs so that the ith arc exits from the crossing P_i. Call u_i the meridian of the ith arc.

Call D the $(k \times k)$matrix where the ith column gives the coordinates of the boundary of a lifting of $D(P_i)$ in the maximal abelian covering of $S^3 \backslash L$ with respect to liftings of the u_i. Call D_{kk} its submatrix obtained by deleting the kth row and the kth column.

Thus, according to the Hartley definition, up to positive units of $\mathbb{Z}[Q_{fa}(S^3 \backslash L)]$,

$$(\exp(u_k) - 1)\Delta_{Hartley}(L) = \det(D_{kk})$$

A.5.C Comparing with the normalization of this appendix

To see that, according to the definition of this appendix,

$$(\exp(u_k) - 1)\Delta(L) = (-1)^{n-1}\det(D_{kk}),$$

it suffices to prove that the cellular decomposition $D(P_1)$, ..., $D(P_{k-1})$, u_1, ..., u_k is "$(-1)^{n-1}$oriented".
We can and do assume that the arcs are numbered so that u_{k-n+i} is an arc of the component K_i for i = 1, ..., n, and so that from 1 to (k-n) we number the arcs successively as we meet them when following the successive components K_i, (i=1, ...,n) according to their orientations from the end of u_{k-n+i} to its beginning.
Perform then the following oriented changes of basis.
Add all two-cells corresponding to crossings where K_i crosses under (except $D(P_{k-n+i})$) to $D(P_{k-n+i})$. The obtained sum represents the tubular neighborhood of K_i in $H_2(S^3\backslash L)$ oriented, contrary to the conventions of A.1.5, as part of the boundary of L's exterior. This will give the "$(-1)^{n-1}$".
Subtract u_{k-n+i} from all the oriented meridians of K_i except from u_{k-n+i}.
Now, to check the "$(-1)^{n-1}$orientation" of the presentation, it suffices to prove that the submatrix D' of D, intersection of the (k-n) first rows and columns of D, is mapped by the augmentation morphism to a matrix $\varepsilon(D')$ with a positive determinant. (Note that the previous changes of basis left D' invariant.)
With the choices above, $\varepsilon(D')$ is diagonal by blocks (a block corresponds to a component of L) and each block has the diagonal of the identity matrix, has -1 as terms just above the diagonal and is zero elsewhere. So the determinant of $\varepsilon(D')$ is 1. This proves 2.3.11.

❑

§A.6 Relations with one-variable Alexander polynomials

Let L be an oriented null-homologous n-component link in an oriented rational homology sphere M. Let $R_c(M\backslash L)$ denote the infinite cyclic covering of $M\backslash L$ obtained from the morphism ϕ_c from $\pi_1(M\backslash L)$ to $\mathbb{Z}$ which sends the oriented meridians of L to 1.
Let $P_L(t)$ be defined from the normalized Alexander polynomial of L (see 2.1.1) as

$$P_L(t) = \begin{cases} \Delta(L)(t) & \text{if } n = 1 \\ \Delta(L)(t, t, \ldots, t)(t^{1/2} - t^{-1/2}) & \text{if } n \geq 2 \end{cases}$$

$H_1(R_c(M\backslash L); \mathbb{Z})$ thus has a natural structure of $\mathbb{Z}[t,t^{-1}]$-module, and P_L is easily seen to be its order, that is, the determinant of one of its square presentation matrices. The only interest of the following proposition is to compare the normalization obtained here for the one-variable polynomial with standard normalizations.

PROPOSITION **A.6.1**:
Let L *be an oriented null-homologous link in an oriented rational homology sphere* M, *and let* Σ *be an oriented Seifert surface of* L. *Let* V *be the matrix of the Seifert form of* Σ *with respect to a basis of* $H_1(\Sigma;\mathbb{Z})$. *Then:*

$$|H_1(M;\mathbb{Z})| \det(t^{1/2} V - t^{-1/2}\, {}^tV) = P_L(t)$$

PROOF: View $P_L(t)$ as defined from $\mathbb{Z}[t,t^{-1}]$-equivariant well-oriented (with respect to $o_L(M\setminus L)$) presentations of deficiency 1 of $H_1(M\setminus L;*;\mathbb{Z}[t,t^{-1}])$, just copying the definition of τ (of §A.2) in the case of this simpler covering. If the last generator of such a presentation is sent to 1 by ϕ_c, $P_L(t)$ is (up to positive units of $\mathbb{Z}[t,t^{-1}]$) the determinant of the matrix D' obtained from the presentation matrix by deleting the row of this last generator; and D' is a presentation matrix of $H_1(M\setminus L;\mathbb{Z}[t,t^{-1}]) = H_1(R_c(M\setminus L);\mathbb{Z})$ with a prescribed relative orientation of its rows and columns.
Keep this in mind when computing $P_L(t)$ with the Seifert surface mentioned above.

(a) A first presentation of $H_1(M\setminus L;\mathbb{Z}[t,t^{-1}])$
Since the morphism ϕ_c associates to a loop of $M\setminus L$ its algebraic intersection number with Σ, $R_c(M\setminus L)$ can be seen as $\mathbb{Z}$ copies of $M\setminus\Sigma$ glued along $\mathbb{Z}$ liftings of $\Sigma\setminus L$.
So, according to the Mayer-Vietoris sequence,

$$H_1(M\setminus L;\mathbb{Z}[t,t^{-1}]) = \frac{H_1(M\setminus\Sigma;\mathbb{Z}) \otimes_{\mathbb{Z}} \mathbb{Z}[t,t^{-1}]}{\bigoplus_{i=1}^{n-1+2g} \mathbb{Z}[t,t^{-1}]\,(t\, e_i^+ - e_i^-)}$$

where $(e_i)_{i=1,\ldots,n-1+2g}$ is a basis of the $H_1(.;\mathbb{Z})$ of a lifting Σ_0 of $\Sigma\setminus L$, and the superscripts + and - denote the push-offs in the positive and negative normal directions of Σ respectively. Choose the basis (e_i) so that e_{i+2g} is homologous to a curve parallel to K_i for i=1, ..., n-1, and the first 2g vectors form a symplectic system for the intersection form on Σ_0.

(b) Specifying a presentation of $H_1(M\setminus\Sigma;\mathbb{Z})$
View a regular neighborhood of Σ in M as a handlebody, regular neighborhood of the wedge of representatives of the e_i's. Let $m(e_i)$ be a meridian of e_i on the boundary of this handlebody. $m(e_{i+2g})$ is homologous to (m_i-m_n). According to the Mayer-Vietoris sequence, $\{m(e_i)\}_{i=1,\ldots,n-1+2g}$ is a basis of $H_1(M\setminus\Sigma;\mathbb{Q})$. Note furthermore that the $m(e_i)$-coordinate of an element x with respect to this basis is the linking number $Lk_M(x,e_i)$.
So, $H_1(M\setminus\Sigma;\mathbb{Z})$ has the following form:

$$H_1(M\setminus\Sigma;\mathbb{Z}) = \bigoplus_{i=1}^{n-1+2g} \mathbb{Z}\, g_i \oplus \bigoplus_{i=1}^{k} \frac{\mathbb{Z}}{n_i\mathbb{Z}} t_i$$

where k is an integer, the n_i's are positive integers and the g_i's and the t_i's are generators of $H_1(M\setminus\Sigma;\mathbb{Z})$.
Write $m(e_j)$, up to a torsion element as

$$m(e_j) = \sum_{i=1}^{n-1+2g} a_{ij}\, g_i$$

and let A be the matrix $A=[a_{ij}]_{1\leq i,j\leq n-1+2g}$. Assume (after possibly changing a g_i into its opposite) that A has a positive determinant.
Note that:

$$|H_1(M;\mathbb{Z})| = \det(A)\prod_{i=1}^{k} n_i$$

(c) Equivariant explicit presentation of $H_1(M\setminus L;\mathbb{Z}[t,t^{-1}])$
Take as generators the $(t_i)_{i=1,\ldots,k}$ and the $(g_i)_{i=1,\ldots,n-1+2g}$ now viewed as homology classes of a preferred lifting of $M\setminus\Sigma$, and as relations $(n_i t_i)_{i=1,\ldots,k}$, $(te_i^+ - e_i^-)_{i=1,\ldots,n-1+2g}$ written as combinations of the g_i's and the t_i's.

The matrix of this presentation is the product $\begin{pmatrix} I & ? \\ 0 & A \end{pmatrix}\begin{pmatrix} N & ? \\ 0 & tV - {}^tV \end{pmatrix}$ where I and N denote the (k×k) diagonal matrices with respective diagonals (1, ..., 1) and $(n_1, \ldots, n_k)$.
Hence, we already see that $P_L(t)$ is equal to $|H_1(M;\mathbb{Z})|\det(tV - {}^tV)$ up to units in $\mathbb{Z}[t^{1/2},t^{-1/2}]$. After symmetrization, A.6.1 is true up to sign.

(d) Checking orientations to get the right sign
Note that for our current purposes, the t_i's can be forgotten and, since det(A) is positive, the g_i's may be replaced by the $m(e_i)$'s.
So, to get the right sign, and hence the complete result, it suffices to check the following statement:
If a lifting g_{n+2g} of an oriented meridian of K_n, were added to the previous generators, the obtained presentation would be a well-oriented presentation of $H_1(M\setminus L;*;\mathbb{R}[t,t^{-1}])$.
This is the consequence of the three following facts:
Fact 1: The augmentation morphism maps the determinant of $[(tV - {}^tV)_{ij}]_{1\leq i,j\leq 2g}$ to 1.
(because $\varepsilon([(tV - {}^tV)_{ij}]_{1\leq i,j\leq 2g}) = [(V - {}^tV)_{ij}]_{1\leq i,j\leq 2g}$ is the matrix of the intersection form of Σ, with respect to $(e_i)_{i=1,\ldots,2g}$)

Fact 2: The n last generators are $(m_1\text{-}m_n)$, $(m_2\text{-}m_n)$, ..., $(m_{n-1}\text{-}m_n)$, m_n (+-equivalent to m_1, ..., m_n).

Fact 3: The (n-1) last relations $(t\, e^+_{i+2g} - e^-_{i+2g})_{i=1,\ ...,\ n-1}$ "are" oriented boundaries of liftings of the oriented boundaries $\partial T(K_i)$, i = 1, ..., n-1. This is shown by the following figure.

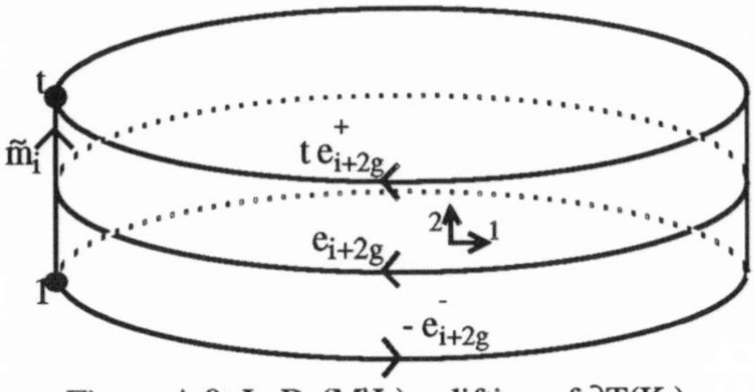

Figure A.9: In $R_c(M\backslash L)$, a lifting of $\partial T(K_i)$

(Σ_0 is outside the pictured cylinder.)

❑

PROOF OF 2.3.14 AND 2.3.15

A polynomial P in $t^{k/2}\mathbb{Q}[t, t^{-1}]$ which satisfies

$$P(t^{1/2}) = (-1)^k P(t^{-1/2})$$

is a polynomial of $\mathbb{Q}[z = (t^{1/2} - t^{-1/2})]$ and the degrees that arise in its expression are each congruent to k (mod 2).

(It can be seen by induction on the degree of P by dividing (P - P(1)) by z.)

The fact that the Conway polynomial ∇_L of an n-component link L is divisible by z^{n-1} can be seen as in the proof of Lemma 5.2.5, or from 2.3.13 and 2.5.2.

❑

PROOF OF THE SKEIN RELATION 2.3.16

Choose first a connected Seifert surface Σ^0 for L^0 which intersects "the" ball B where L^0, L^+ and L^- are different as follows:

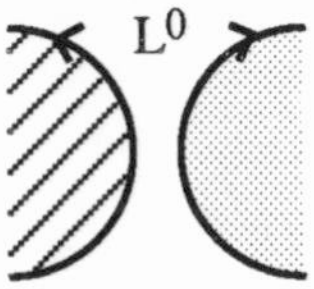

Figure A.10

Construct Σ^+ and Σ^- from Σ^0 so that Σ^0, Σ^+ and Σ^- are identical outside B, and, Σ^+ and Σ^- intersect B as in Figure A.11:

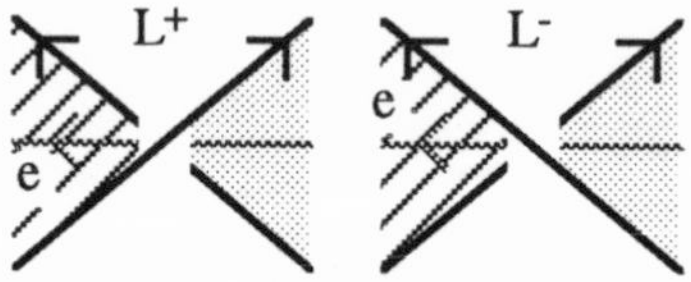

Figure A.11

Then, if e is a loop of Σ^+ and Σ^- intersecting B as in Figure A.11,

$$H_1(\Sigma^+;\mathbb{Z}) \cong H_1(\Sigma^-;\mathbb{Z}) = \mathbb{Z}e \oplus H_1(\Sigma^0;\mathbb{Z})$$

Let e_+^+ and e_-^+ respectively denote the push-offs of e in the positive normal direction to Σ^+ and to Σ^-, it is now clear from 2.3.14 that

$$\nabla_{L^+} - \nabla_{L^-} = z(\,Lk(e_+^+,e) - Lk(e_-^+,e)\,)\nabla_{L^0}$$

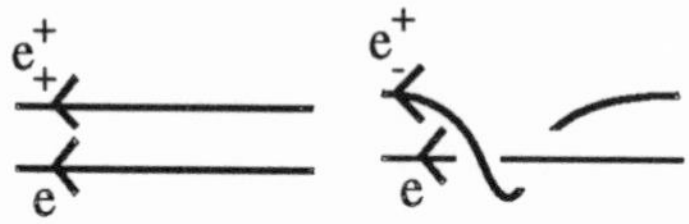

Figure A.12

Figure A.12 shows e_+^+ and e_-^+ in B and proves:

$$Lk(e_+^+,e) - Lk(e_-^+,e) = -1$$

This completes the proof of 2.3.16.

❑

Bibliography

[A-M] S. AKBULUT and J. MCCARTHY: *Casson's invariant for oriented homology 3-spheres, an exposition,* Mathematical Notes 36, Princeton University Press, 1990.

[B-L 1] S. BOYER and D. LINES: *Surgery Formulae for Casson's invariant and extensions to homology lens spaces,* Journal für die reine und angewandte Mathematik, Vol. 405, 181-220, 1990.

[B-L 2] S. BOYER and D. LINES: *Conway potential functions for links in* $\mathbb{Q}$*-homology 3-spheres,* Proceedings of the Edinburgh Mathematical Society, Vol. 35, 53-69, 1992.

[C-F] R. H. CROWELL and R. H. FOX: *Introduction to knot theory,* Ginn, 1963.

[F-M-S] S. FUKUHARA, Y. MATSUMOTO and K. SAKAMOTO: *Casson's invariant of Seifert homology 3-spheres,* Math. Ann., Vol. 287, 275-285, 1990.

[F-R] R. FENN and C. ROURKE: *On Kirby's calculus of links,* Topology, Vol.18, 1-15, 1979.

[G-M 1] L. GUILLOU and A. MARIN: *A la recherche de la topologie perdue,* Progress in Math., Vol. 62, Birkhäuser, 1986.

[G-M 2] L. GUILLOU and A. MARIN: *Notes sur l'invariant de Casson des sphères d'homologie de dimension 3,* L'Enseignement Math., 38, 233-290, 1992.

[Hart] R. HARTLEY: *The Conway potential function for links,* Comment. Math. Helvetici, Vol. 58, 365-378, 1983.

[Hos] J. HOSTE: *A formula for Casson's invariant,* Trans. A. M. S., Vol. 297, n°2, 547-562, 1986.

[Ka] S. KAPLAN: *Constructing framed 4-manifolds with given almost framed boundaries,* Trans. A.M.S., Vol. 254, 237-263, 1979.

[Kir 1] R. C. KIRBY: *The topology of 4-manifolds,* Lecture Notes in Math., Vol. 1374, Springer-Verlag, 1989.

[Kir 2] R. C. KIRBY: *A calculus for framed links in* S^3, Inventiones math., Vol. 45, 35-56, 1978.

[L 1] C. LESCOP: *Invariant de Casson-Walker des sphères d'homologie rationnelle fibrées de Seifert,* Notes aux C. R. Acad. Sci. Paris, t. 310, Série I, 727-730, 1990.

[L 2] C. LESCOP: *Un Algorithme pour calculer l'Invariant de Walker,* Bull. Soc. math. France, 118, 363-376, 1990.

[L 3] C. LESCOP: *Sur l'invariant de Casson-Walker: Formule de chirurgie globale et généralisation aux variétés de dimension 3 fermées orientées,* C. R. Acad. Sci. Paris, t. 315, Série I, 437-440, 1992.

[Marin] A. MARIN: *Un nouvel invariant pour les sphères d'homologie de dimension 3 (d'après Casson),* Séminaire Bourbaki n° 693, 1988.

[Mil] J. MILNOR: *Spin structures on manifolds,* L'Enseignement Math., 9, 198-203, 1963.

[Mo] J. M. MONTESINOS: *Classical Tessellations and Three-Manifolds,* Universitext, Springer-Verlag, 1987.

[N-W] W. NEUMANN and J. WAHL: *Casson invariant of links of singularities,* Comm. Math. Helv., Vol. 65, 58-78, 1990.

[R-G] H. RADEMACHER and E. GROSSWALD: *Dedekind sums,* The Carus Mathematical Monographs n°16, 1972.

[Ro 1] D. ROLFSEN: *Knots and links,* Publish or Perish, 1976.

[Ro 2] D. ROLFSEN: *Rational surgery calculus: Extension of Kirby's theorem,* Pac. Journal of Maths, Vol. 110, n°2, 1984.

[Rou] C. ROURKE: *A new proof that Ω_3 is zero,* J. London Math. Soc. (2) 31, 373-376, 1985.

[Seif] H. SEIFERT: *Topology of 3-dimensional fibered spaces,* A Textbook of Topology (H. Seifert & W. Threlfall), Academic Press 1980, 359-422, English translation of : *Topologie dreidimensionaler gefaserter Raüme,* Acta Mathematica 60, 147-288, 1933.

[Tu] V. G. TURAEV: *Reidemeister torsion in knot theory,* Russian Math. Surveys, 41:1, 119-182, 1986.

[W] K. WALKER: *An Extension of Casson's Invariant,* Annals of Mathematics Studies, 126, Princeton University Press, 1992.

Christine Lescop, Université de Grenoble I, Institut Fourier, Laboratoire de Mathématiques associé au CNRS (URA 188), B. P. 74, 38402 Saint Martin d'Hères Cédex (FRANCE)

e-mail address: lescop@puccini.ujf-grenoble.fr

Index